The
Development
of
Newtonian Optics
in
England

The Development *of* Newtonian Optics *in* England

Henry John Steffens
University of Vermont

Science History Publications/USA
New York
1977

QC
352
S 73

Science History Publications/USA
a division of
Neale Watson Academic Publications, Inc.
156 Fifth Avenue
New York, New York 10010

© Neale Watson Academic Publications, Inc. 1977

Designed and manufactured in the U.S.A.

Library of Congress Cataloging in Publication Data
Steffens, Henry John.
    The development of Newtonian optics in England.
    Bibliography: p.
    Includes index.
    1. Optics—History—England.   I. Title.
QC352.S73   535′.0942   76-2493
ISBN 0-88202-048-X

Contents

	Preface	vii
I	Sir Isaac Newton and the Newtonians	1
	Social and Philosophical Backgrounds	*1*
	Sir Isaac Newton's Science	*10*
	The Beginnings of Newtonian Optics	*27*
II	The Construction of a System	55
	Inheritors of Forces and Corpuscles	*55*
	The Rival Undulatory Theory in England	*92*
III	Thomas Young and the Challenge to Newtonian Optics	107
	Thomas Young's Concept of Interference	*107*
	Henry Brougham's Attack	*128*
IV	The Tenacity of Newtonian Optics in England	137
	David Brewster, the Last Champion	*137*
	Epilogue	*149*
	References	153
	Selected Bibliography	177
	Index	189

Preface

This history of Newtonian optics in England has been presented in the form of a continuous development to give the study more coherence and clarity. Rather extensive quotations from primary sources have been included where appropriate. This was done for two reasons: to avoid the problems of generality and narrative statement by providing documentation, and to allow the reader to formulate his own opinion of the author's interpretation of scientific contributions.

The study of a particular system of optics is, of necessity, of a narrower scope than a general history of physical optics during the eighteenth and early nineteenth centuries. Attention has been concentrated upon those men, events, discoveries, and concepts which shaped the development of a specific system of ideas in physical optics, called Newtonian optics. This system of ideas has been treated in many respects as if it possessed a continuing life of its own in the eighteenth century. Historical evidence shows clearly that natural philosophers in the eighteenth and early nineteenth centuries pursued their investigations of physical optics as if there were a definable system of Newtonian optics. Certainly by the 1730s, there had emerged in England a way of thinking about physical optics which was coherent enough to be supported as a unified system of ideas.

The development of the wave theory of light has been included in this study only in its early stages, as it directly affected the development of Newtonian optics in England. The history of the wave theory, especially its development after the conception of the transverse wave by Thomas Young and Augustin Fresnel, was outside the scope of this work. A history of the wave theory in the nineteenth century needs to be written, but it deserves an extended treatment of its own.

The distinction between Newton's optics and Newtonian optics has been maintained throughout this work. Chapter I is intended to distinguish the basic reasons for the substantial difference between the two. The history of Newtonian optics provides an excellent example of how the works of a master can be changed and shaped to meet the needs of his disciples. It also points out in detail the fallacy of believing that Newton's science, or anyone else's, is objective, free from metaphysical assumptions, and exclusively empirical. The empirical facade of Newtonian optics served only to disguise concurrent assumptions about light and about the nature of the physical world.

This work has arisen directly from interests begun during the preparation of my master's thesis, of the same title, at Cornell University in 1965. I have, of course, expanded and extended my work, but the basic conclusions and topics remain the same. Since that time, two excellent books have appeared which supplement my work: A.I. Sabra's *Theories of Light from Descartes to Newton* (London, 1967) and Robert E. Schofield's *Mechanism and Materialism, British Natural Philosophy in the Age of Reason* (Princeton, 1969). Peter Anton Pav's unpublished Indiana University doctoral dissertation, 1964, *Eighteenth-Century Optics: The Age of Unenlightenment* (University Microfilms, Inc., 65-3510) can also be read with profit for a different perspective toward eighteenth century optics than mine, and for his treatment of French optical works in the eighteenth century. Three other unpublished doctoral dissertations supplement my work and have sections covering similar ground: D.B. Wilson, *The Reception of the Wave Theory of Light by Cambridge Physicists,* The Johns Hopkins University, 1968; R.H. Silliman, *Augustin Fresnel and the Establishment of the Wave Theory of Light,* Princeton University, 1968; and E.W. Morse, *Natural Philosophy, Hypotheses and Impiety: Sir David Brewster Confronts the Undulatory Theory of Light,* University of California, Berkeley, 1972.

I would like to extend my appreciation to Professor Henry Guerlac, who kindled my interest in the history of optics in his seminar by his knowledge of Newton's science and by his suggestion that I investigate the works of Robert Smith. My sincere thanks go to Professor L. Pearce Williams for his guidance throughout my graduate years and for his continued encouragment and constructive criticism.

Special thanks are due my wife, Judy, for her help and patience as I pursue my studies.

Burlington, Vermont

I
Sir Isaac Newton and the Newtonians

Social and Philosophical Backgrounds

Sir Isaac Newton lived in a world which was changing rapidly and fundamentally. He contributed his small share to changes in the English society of his time, and, of course, his work and influence served to reinforce a dramatic alteration in the philosophical and scientific outlook of the western world. His contributions to the development of modern thought are difficult to overestimate. It is interesting to note, however, that Sir Isaac would have personally rejected or seriously disagreed with many of the developments in both science and philosophy put forward in his name. The history of Newtonian optics in England offers an instructive example of how well-meaning disciples and changes in societal conditions can alter and restructure the works of a great man.

During the eighteenth century a system of optics was developed which was based on the assumption of the corpuscularity of light and the mutual forces of attraction and repulsion which acted between light and matter. Called Newtonian optics, this system can not be attributed directly to Sir Isaac Newton without very careful qualification. The Newtonian system was quite different, in the most fundamental ways, from Newton's own optical works. Newton differed from his enthusiastic followers in his attitudes toward corpuscularity, toward forces operative between light and matter, and most especially on questions of interpretation of the science which could be seen to impinge upon questions of the nature of the Deity and of God's activity in the world. Newton also had his own distinct ideas on what may now be called scientific method. His cautionary strictures against hypotheses and his stress on what he called experimental philosophy were only imperfectly understood by his disciples. With few exceptions, Newton's followers captured the letter of his scientific method while they failed to appreciate the spirit. This is simply to say that Newton was both a man of the seventeenth century and a man of the new, so-called Age of Reason. The meaning of many of his most basic assumptions and beliefs could be found only in the context of a century still concerned with the struggle between the "ancients and the moderns" and with the expunging of Aristotelian and Scholastic notions from all forms of thought. To many of Newton's disciples in the eighteenth century, Newton's own deep concerns over questions of the nature of matter, and force, and over God's activity in the world lacked immediacy and were frequently dismissed as unimportant. Newtonian optics was obviously derived from Newton's work in optics, but the emphases and the assumptions which underlay both were fundamentally different in kind. Newton was still very much a man of the seventeenth century, although, as we shall see, he managed to extricate himself sufficiently from the older modes of thought to provide the source and inspiration for much of the scientific work of the eighteenth century.

The seventeenth century saw a change in the way in which western men viewed the physical world. Like all changes in intellectual orientation, this changing world-view included many discernible contributions from the past. However, it also represented a new way of perceiving and describing old problems. The first half of the seventeenth century saw a turning away from the time-honored modes of Aristotelian-Scholastic thought toward forms which stressed physical explanations, mathematics, and practical demonstrations. The Scholastic emphasis on the "qualities" and the stress on "causal," philosophical explanations were losing ground to a concentration on the revival of ancient atomism and modes of thought which sought to provide physical explanations for all phenomena. The reasons for this change in orientation are numerous and sometimes seem intangible, but the evidence that there was such a shift is conclusive.[1]

The seventeenth century was close enough to the middle ages in background and education so that medieval modes of thought persisted and were evident in the works of the seventeenth century natural philosophers. But these modes of Scholastic thought were confronted at every turn by a newly forged alliance between rational speculation and atomistic concepts. The major questions which spurred scientific controversy in the seventeenth century tended to be, at their foundation, questions of philosophical orientation. It was a century which saw the abandonment of teleological and strictly philosophical explanations in favor of explanations which involved mechanisms of matter and motion. The interpretation of God's place in the physical world also changed dramatically, along with this new emphasis. Nature was no longer deduced from the consideration of final causes and the attributes of God. Natural philosophers focused their attention upon the laws which seemed to operate in nature. They then began to see proof for the existence of God in their descriptions of the regularities of nature. God lost His place as the first concern of intellectual endeavor. But He was remembered with enthusiasm, once nature had been examined. God still had the important position of being responsible for the observed activities in the world. But He was now the second consideration, after nature had been investigated with sufficient care to indicate evidence for His existence and activity.

The emphasis upon the world was an attribute of the age, not only an aspect of natural philosophy. The overseas voyages, the changing European economy, the emerging structures of national states, the realities of religious warfare, the growing governmental awareness that science could be useful, all indicated a new emphasis on the events of the world.[2] But the most important aspect of the early seventeenth century for the development of natural philosophy was the combination of increased reliance upon rational speculation and the revived interest in ancient atomism.[3]

The seventeenth century enthusiasm for atomism was derived from a renewed interest in the writings of such ancient atomists as Democritus, Epicurus, Lucretius, and Hero of Alexandria. The reasons behind this renewed interest are numerous and varied.[4] Aristotelian physics was no longer capable of providing explanations which matched the temper of the times. The reliance upon substantial forms and qualities did not provide the kind of immediate statement related to experience that the men of the seventeenth century were coming to expect. The Aristotelian-Scholastic arguments increasingly seemed to be obscure collections of words when contrasted with explanations drawn from common experience. Natural philosophers after 1600 were increasingly familiar with the analogy of the machine and mechanical explanations drawn from elements of their experience. They were more familiar with the crafts, with the presence of mechanical devices and machines in their daily lives, and they were better trained in mathematics. The means of explanation possible within the framework of atomism held out increasing appeal. The world could be explained in terms of matter and motion, in terms of the sizes and shapes of particles of matter and their combinations and motions which produced all the phenomena of the world. The impact of matter upon matter produced motion and change. When the mechanism which produced such motion and change was obscure, as in the case of the lodestone and magnetic phenomena, atomism provided the easy and persuasive argument of subvisible effect. Turning to the ancients for possible modes of expression was also in keeping with the widespread rejection of Scholasticism and Aristotelian authority. The ancient atomists were certainly outside of this tradition. Often they also had the advantage of expressing themselves as did Lucretius, in classical Latin, which accorded well with the humanist emphasis on reading and consideration of classical texts.

The fact that atomism provided an alternative to Aristotle may well have been the decisive factor. Aristotelian physics, with its emphasis on teleology and the organic analogy, was increasingly seen by men of the sixteenth and early seventeenth centuries as inadequate to meet the specialized questions about local motion in physics, about certain chemical problems and the new Torricellian vacuum experiments.[5] Aristotelian physics was viewed as having serious problems in its own right, before many philosophical arguments were brought to bear.[6] Aristotelian physics could not accommodate mechanical interpretations because it was concerned with causes, not descriptions of events. The seventeenth century felt less and less comfortable with Aristotle's "animated" or "organic" analogies, applied to matter and final causation. The concepts of the Unmoved First Mover and the modes of action which produced motion all involved the human analogies of desire and love. This analogy was clearest in the explanation of the motion of the heavenly bodies. "The heavenly spheres, then, move by Eros, by desire to achieve as far as possible the perfect good of the unmoved mover, and this they do by executing the most perfect motion, namely, uniform circular motion. Love, in fact, makes the world go round. This is not the only occasion in Aristotle's philosophy on which he introduces love or desire as an explanatory cause; it is one of his fundamental analogies for final causation. He speaks, as he says 'metaphorically' of the potentiality of matter for acquiring form as desire: matter desires form 'as the female desires the male, and the ugly the beautiful' and all motion to the natural place is actuated by the desire of the potential to become actual, and therefore involves, on the level of final causation, attraction at a distance."[7]

Once the Aristotelian notions of causation and the organic analogies began to come under close scrutiny, especially in the cases of the mechanics of local motion and celestial motions, they were found to be inadequate. In this sense, Aristotelian physics contained the seeds of its own destruction in having called close attention to the phenomena of local motion. The first questionings of the applicability of Aristotelian concepts to problems like the motions of falling bodies and spinning tops produced the medieval corpus of works which led to theoretical mechanics. But the abandonment of the organic analogy gave incentive to the formulation of a new analogy or conceptual scheme to accompany the emerging critical thoughts in mechanics.[8] The new analogy was the mechanical analogy, a conceptual framework which suited both the seventeenth century orientation and the revival of the ancient system which had always been separate from Aristotelian physics, ancient atomism.

The revived interest in atomism involved more than the restatement of the ideals of the ancients. The new atomism had to find its place in the context of the new emphasis on the mechanical analogy and of the new experiments on motion, the new Copernican astronomy, experiments in chemistry, and experiments with the barometer and air pressure. It was an ancient conceptual scheme in a new societal environment, placed within an expanded context of observation and experience. This revival of atomism has been described by Robert Kargon as follows:

> The new atomic doctrine had to incorporate within it the experimental advances for the sake of which, in part, it was revived. At first the atomic theory was used to solve particular difficulties in the explanation of natural phenomena, as for example, in the early work of Thomas Harriot. Some, like Daniel Sennert and Nicholas Hill, tried to bridge the gap between atomism and Aristotelianism with attempts at synthesis. Finally, others like Pierre Gassendi and the corpuscularians Thomas Hobbes and René Descartes built great entire systems, going far beyond their classical predecessors in wealth of detail.
>
> The history of atomism as part of the establishment of the mechanical philosophy provides an excellent case study in the transition from one world-view to another. The atomic hypothesis was, moreover, viewed by many early scientists of the seventeenth century as a vital part of their work. The greatest scientific minds of the age were engaged in an elaboration and modernization of the doctrine.[9]

Sir Isaac Newton's early work in optics began in this scientific environment of the elaboration and application of the new atomic doctrine, within the more complete conceptual framework called the mechanical philosophy. As we shall see, his early optical writings are quite closely allied to the assumptions about the physical world espoused by men such as Robert Boyle and Isaac Barrow, who took atomism very seriously. Newton's early atomism was a composite of current atomist assumptions and a concern for the religious implications of those assumptions.

Charges of atheism had always dogged the steps of the atomists. The seventeenth century had its full measure of fear of the implications of explanations of the physical world in terms of matter, motion, and the void. Thomas Harriot, one of the most important of the early supporters of atomism in England in the opening years of that century, was reluctant to advocate openly the doctrines of matter and the void because of charges of atheism. But it is clear that atomism had an important place in shaping Harriot's work in natural philosophy.[10] Thomas Harriot may be seen as an excellent example of the transition to a new world view in the seventeenth century. An interesting comparison can be made between Johannes Kepler and Thomas Harriot in this regard. Harriot was able to bring a new interpretation to his interest in optical phenomena because of his acceptance of atomism. "Whereas Kepler, for instance, clung to Aristotelian modes of scientific explanation, Harriot liberated himself from the scholasticism in which he was trained, and turned to ancient atomism for his basic principles. For Kepler, the nature of reflection and refraction could be explained in a satisfying manner by invoking two contrary qualities, translucence and opacity. For Harriot, this treatment lacked vigour and clarity. He required what can be termed a *physical* rather than a *verbal* explanation. The basis for his view was the teachings of the ancient pagan philosophers.

"In two very real and important senses, Harriot's was a *mechanical* philosophy. First, the widespread introduction of mining and other machinery provided both a convenient physical analogy for natural philosophy and an important subject for explanation. Harriot and his friends used the analogy of the machine to explain phenomena, and in turn explained the action of machines by matter and motion in the void . . . . Secondly, Harriot and his disciples relied solely upon matter and its motion for physical explanation. It is this unique reliance which marked the mechanical philosophies of the mid-seventeenth century."[11]

By mid-century, it was clear that the mode of thought using the machine analogy, the "mechanical philosophy," was providing the most viable way of viewing the physical world. This was true of the biological as well as the physical sciences. This mode of thought was especially well developed, although differently, by René Descartes and Pierre Gassendi. Both these men formulated systems of thought which had the dual purpose of replacing the old Aristotelian-Scholastic tradition with a "new philosophy" and of attempting to explain the world in terms of matter and motion. But it was also recognized by mid-seventeenth century that this mode of thought had to surmount two grave problems: the charge of atheism and the charge that the grand schemes of these mechanical philosophers were reasonable and plausible, but were only hypothetical.

The fear of atheism in England was exacerbated by the turmoil of the times and by the work of Thomas Hobbes. After Elizabeth I died in 1603, England experienced a growing political and religious tension under her two Stuart successors, James I and Charles I. The masterful Elizabeth had established a working relationship with Parliament through a combination of diplomacy and cunning, but James I proclaimed his firm belief in the divine right of kings. James set both nobles and gentry interested in constitutional government against him at a time when support for Parliament was becoming increasingly effective. The division widened under the rule of his son, Charles I, developing into a serious constitutional struggle. To this conflict the Puritans added increasingly vocal criticism of the English Church. Civil War broke out in England and resulted in the execution of Charles I and the attempt to establish a republican form of government from 1649-1660, under the leadership of Oliver Cromwell and of his son Richard. Cromwell's death in 1658 weakened the movement sufficiently to allow the restoration of Charles II to the English throne in 1660. Charles recognized both the need for caution and the tenuous nature of his royal power, but his death in 1685 brought his secretly Catholic brother, James II, to the throne. James brought the constitutional struggle to a crisis through problems caused by his religion and his inept political maneuverings. The Glorious Revolution of 1688 finally established a constitutional basis for the English monarchy. England emerged from these struggles with a constitutional government which proved extremely stable and which provided the basis for the growth of English wealth and naval supremacy in the eighteenth century. England, after Newton died in 1727, was a very different place from the country into which he was born in 1642.

Thomas Hobbes introduced an element of philosophical turmoil into this troubled England of the early 1650s. Hobbes had become a close friend of Pierre Gassendi during the 1640s when his circle of friends, the Newcastle Circle, moved to Paris.[12] Hobbes was quite taken with Gassendi's atomism, particularly as expressed in the *Animadversiones in decimum librum Diogenis Laertii* of 1649. Hobbes drew from his work some startling theological conclusions, especially in relation to the notion of spirit and of soul. Hobbes denied the presence of immaterial spirits in the world. He wanted to be rid, once and for all, of the notion of an immaterial spirit as a causal agent in the world. Hobbes believed that the reliance upon such immaterial (and therefore immortal) entities amounted to a reliance upon occult qualities. These occult qualities should have no place in natural philosophy. In *Of Human Nature* (written in 1640, published 1650), Hobbes wrote:

"'By the name *spirit*, we understand a *body natural*, but of such subtility that it worketh not upon the senses; but that filleth up the place which the image of a visible body might fill up.' That there exist *immaterial* spirits is a flat contradiction in terms, for a spirit is a substance, and all substances have dimension and corporeality."[13]

Hobbes called the immateriality and immortality of the soul into serious question, and by logical extension suggested that God was a substance and therefore corporeal. Hobbes insisted upon the materiality of all spirits; if they were anything, they were material and man's body was not inhabited by them. His unorthodox position prompted the bishops in Parliament to wish Hobbes burnt as a heretic. He was variously attacked "as an Epicurean, atomist, Anthropomorphist, Sadducean, Manichean, Luciferian, and Jew."[14]

Hobbes' staunch assertion of the materiality of spirits caused serious embarrassment for the English supporters of atomism after 1650. Most atomists were not willing to agree with the expulsion of immaterial spirits from the universe. Amongst those who would not follow Hobbes in this regard was Isaac Newton. The question of the existence and activity of immaterial spirits became an important and continuing question for the English. It is only in this context that Newton's references to "active principles" and "subtile matter" take on any full meaning. Newton believed that to maintain that the soul was material was to be atheistical. The broad problem, with which Newton remained deeply concerned, was the explanation of how immaterial agents and entities could affect the material world of matter and motion. If God was to be considered active in the world, as in Newton's view He must be so considered, then how was His action in the world possible and what was the possible means for this activity?

These religious questions lost their immediacy and importance to the generation after Newton. As we shall see, this explains why Newton's disciples in physical optics largely neglected all of Newton's serious qualifications of atomism and developed a straightforward system of optics based upon light corpuscles and the attractions and repulsions between light and matter.

Newton's disciples generally failed to understand the depth of his concern for religious questions.[15] They also failed to appreciate the spirit behind his rejection of hypotheses in natural philosophy. Newton sought a method of reasoning in natural philosophy which combined several aspects of the seventeenth century "Scientific Revolution" which had not been successfully united before. He succeeded in combining his conviction that natural philosophy should possess certainty with his skill as an experimenter and his ability as a mathematician. In so doing, Newton created a way of doing science which became the model for the eighteenth and nineteenth centuries. But Newton's own science was firmly placed within the context of his religious beliefs and was, therefore, fundamentally different from the mechanistic, deterministic science so frequently termed Newtonian. All the ingredients of Newton's science, the urge for certainty, the use of experiments, the mathematics, and the religious beliefs, were drawn from Newton's predecessors. His use of them was unique, so unique in fact that his Newtonian successors only accepted part of the combination he synthesized.

The search for certainty in natural philosophy was well under way before Newton began his scientific career. The Royal Society of London was founded in 1662, in large measure to serve this goal of introducing certainty into the study of the physical world. Its members wanted to do away with hypothetical schemes not rooted in experience. The source of this urge for certainty has been located in the works of Francis Bacon. The Royal Society was certainly Baconian, but the diversity of membership was such that the title "Baconian" is difficult to interpret. The term becomes meaningful if we view the link among the Baconian members of the Royal Society as their common dedication to the goal of certainty in natural philosophy through experience.

"Bacon, it will be recalled, proposed to establish successive stages of certainty in science through his new organon. He criticized severely those natural philosophers 'who have taken upon them to lay down the law of nature as a thing already searched out and understood,' for these philosophers, 'whether they have spoken in simple assurances or professional affectation, have done philosophy and the sciences a great injury.' The new organon, or method, on the other hand makes no such claims. It would establish certainty by relying on the evidence of the senses, corrected and aided by safeguards, but rejecting 'the mental operation which follows the act of sense.' Instead of that operation, Bacon would 'open and lay out a new and certain path for the mind to proceed in.' He advocated a path to understanding, 'direct from the sense, by a course of experiment orderly conducted and well built up.' Bacon was, however, aware of the danger of blind empiricism."[16]

Viewed from this standpoint of experience and certainty through reliance upon the senses, the systems of interpretation of the world propounded by René Descartes and Pierre Gassendi were seen to be hypothetical. They were systems devised to replace the Aristotelian scheme, but each was viewed as suffering from the same shortcoming: all were hypothetical. They were logical, plausible, and persuasive, but they could not be viewed as certain. This lack of certainty was appreciated by the mechanical philosophers themselves. In 1636, Hobbes realized:

"In thinges that are not demonstrable, of which kind is the greatest part of naturall philosophy, as depending upon the motion of bodies so subtile as they are invisible, such as ayre and spirits, the most that can be atteyned unto is to have such opinions, as no certayne experience can confute, and from which can be deduced by lawfull argumentation no absurdity."[17]

René Descartes was aware of the same difficulties with the mechanical philosophy. He recognized that the natural philosopher could do little more than construct plausible hypotheses about the world. In a letter to Mersenne in 1638, Descartes wrote:

"To require of me geometrical demonstrations in a question which concerns physics is to ask me to do the impossible . . . . [In such matters we have to rely on suppositions] which, even if they are not exactly true, are yet not manifestly contrary to experience, and in speaking of which we argue consistently, without falling into paralogisms . . . . Take it therefore, that there are only two ways of refuting what I have written, either that of proving by certain experiences or reasons that the things I have supposed are false, or else of showing that what I have deduced from them cannot be so deduced."[18]

Sir Isaac Newton's Science

By the 1660s, there was a general awareness that natural philosophy, as currently practiced, was hypothetical. Using the hindsight available to the historian, we see that this was an entirely appropriate conclusion to draw. Natural philosophy, concerned with the explanation and description of the physical world, was and is still hypothetical. All attempts at objectivity involving the removal of subjective, "non-scientific" elements from science have eventually been recognized as unsuccessful, after the first blush of enthusiasm has waned.[19] The English embarked upon an enthusiastic effort to introduce certainty into natural philosophy in the seventeenth century. This effort received sanction and support from the Royal Society in the early 1660s. In general, there were three lines of approach to certainty, of differing levels of subtlety. There was the attempt to reject all hypotheses whatsoever, and to fall back upon tables and lists of data and observations, claiming to use no theories at all. There was the attempt to put the systems of Descartes and Gassendi to experimental test. Robert Boyle was the most serious and persistent of this group, seeking to discover whether the grand schemes of nature could be seen to conform to experience. But the most important of the seekers after certainty attempted to use experiment and the long neglected area of mathematical demonstration. They attempted to do away with hypothetical physics by the use of experimental test and rigorous mathematics. The primary adherent to this twofold approach was Isaac Newton. He drew his example of experiment mainly from Robert Boyle and his reliance upon mathematics from his teacher Isaac Barrow. Newton was careful to place these two contributions in the context of metaphysical beliefs very closely akin to those held by the Cambridge Platonists, and especially by Henry More.[20]

The thought of the Cambridge Platonists stood in contrast to the work of Thomas Hobbes, in terms of religious implications. Henry More could not accept the removal of spirit from the world, nor the denial that there could be "hidden forces" in nature which were not ascribable to impact or contact forces. More, in short, sought to retain a place for spirit in natural philosophy. His two works, *Immortality of the Soul* and *Enchiridion Metaphysicum,* give rather detailed attention to the actual place of spirit in the operations of the world.[21] More agreed with many of the atomists' assumptions, but he always modified them when they seemed to exclude the possibility of non-corporeal activity. Matter was considered to be composed "of homogeneous atoms, impenetrable as regards each other, without figure, though extended, filling all space, and by their own nature inert, though movable by spirit."[22]

More would not accept any notion of matter possessing innate motive properties. Spirit was necessary, in More's conception of things, to supply motion. He could not specify just what the nature of this spirit was, but he was certain that it must exist. He was also convinced that spirit should have extension, since he believed that all things which exist must be extended. Spirit must therefore be extended, although that was where its similarity to matter ended. Spirit was freely penetrable, and it was able to penetrate matter and to impart motion to matter. It also had powers of contraction and dilation, making it capable of occupying greater or lesser space at will. More summarized his doctrine of spirit in a letter to Robert Boyle, in 1665:

the phenomena of the world cannot be solved merely mechanically, but that there is the necessity of the assistance of a substance distinct from matter, that is, of a spirit, or being incorporeal.[23]

More was convinced that incorporeal spirit was needed for philosophical and theological reasons. He was also persuaded that the natural philosophy of his time had demonstrated conclusively that nature must be more than a simple machine, as some mechanical philosophers maintained. He believed that studies of men had shown the need for incorporeal spirits and he was sure that the continued investigation of magnetism, cohesion, gravity, and chemistry would also support the need for spirits. Since he was confident that mechanical explanations would not suffice, the forces indicated by these phenomena must be other than mechanical, they must be spiritual. This belief in spirits in nature was easily reconcilable with current natural philosophy, but it called for an increase in the level of awareness of the natural philosopher.

Nor . . . needs the acknowledgement of this principle to dampen our endeavors in the search of the mechanical causes of the phenomena of nature, but rather make us more circumspect to distinguish what is the result of the mere mechanical powers of matter and motion, and what of a higher principle. For questionless this secure presumption in some, that there is nothing but matter in the world, has emboldened them too rashly to venture on mechanical solutions where they would not hold.[24]

More was deeply religious, and interested in more than arguments for the existence of spirits. He was concerned with the use of arguments drawn from natural philosophy to support evidence for the existence of God. More was convinced that all of his arguments denying the simple mechanical nature of man and the world implied that there must be not only a spirit of nature but a higher-order incorporeal substance, "a spiritual substance, rational, purposive, supremely worthy of obedience and worship.

'We have discovered out of the simple phenomenon of motion [More's consideration of matter in motion and its ultimate causes] the necessity of the existence of some incorporeal essence distinct from the matter. But there is a further assurance of this truth, from the consideration of the order and admirable effect of this motion in the world. Suppost matter could move itself, could mere matter, with self-motion, amount to that admirable wise contrivance of things which we see in the world? Can a blind impetus produce such effects, with that accuracy and constancy, that the more wise a man is, the more he will be assured, that no wisdom can add, take away, or alter anything in the works of nature, whereby they may be bettered? How can that therefore which has not so much as sense, arise to the effects of the highest reason or intellect?'

More [was] convinced by such teleological proofs that there exists a supremely wise creator and governor of the universe, whose agent and subordinate medium in the execution of his purposes is this lower incorporeal being, the spirit of nature."[25]

Henry More represented a major concern in English religious and philosophical thought in the last half of the seventeenth century. He was fully aware of the new works in natural philosophy, and saw these works, properly interpreted, as support for the denial of atheism and the assertion of the importance of the activity of incorporeal entities in the world. This was a period which was very careful to avoid the materialist implications of atomism. But in doing this, men who adopted More's orientation left themselves open to the charge of re-introducing "occult qualities" into natural philosophy. This stress upon incorporeal spirits merged with the urge for certainty in natural philosophy in a highly interesting way. Sir Isaac Newton was one of the first men to appreciate that experiments could reveal information about incorporeal activity in nature. These experiments should be conducted both to discover truths about the physical world from the phenomena, and also to reveal more completely the mode of God's activity in the world, either directly or through the intermediary of spirits and "active principles." These truths should be expressed mathematically, to reveal the activities in the most accurate way. Experiment and mathematics were combined in Newton's science for the purpose of knowledge of the world, but, more importantly, for the understanding of God's constant activity in the world.

The major aspect of Newton's work which his Newtonian successors either failed to appreciate or consciously dismissed as inappropriate was the importance of Newton's religion to his natural philosophy. The foundation for the differences between Newton's science and Newtonian science was this religious separation. The history of the development of Newtonian optics will make this distinction clear.

Newton believed in God's continual activity in the world. There was no possibility, for Newton, of a mechanical world of matter with innate properties and eternal motion. Such a world could not exist, because it would imply that the world could continue to operate quite apart from God's involvement. This was impossible, Newton believed, because God did act in the world. Newton's maintenance of a continually acting God was much more than an isolated preference in natural philosophy. It was supported by the whole of his theological beliefs and his attitude toward history. Newton believed in a transcendent, Christian God who worked through history. His religious beliefs were ultimately based upon revealed truth, not upon reason or speculation. One of the main sources for revealed truth was the Holy Scriptures. Newton considered the Bible to be true, historical, and entirely reliable. The Bible was factual, not allegorical; the revelations in the Bible were true, and they exactly coincided with the truths which could be recognized from the study of history and the study of natural philosophy. The Bible was a revelation from God, but God also acted and spoke in history. For Newton, there was no conflict between revealed truth and natural philosophy. God's continual activity in the world insured that the study of history and natural philosophy, properly conducted and understood, would coincide exactly with those truths revealed by God in the Scriptures. Newton devoted a great deal of his time to the study of Biblical texts, church history, the chronology in the Old Testament, and commentaries on the prophecies in the Bible.[26] He was as serious a religious scholar as he was a natural philosopher.

Newton's belief in God was indicated to his friends, particularly Robert Boyle and Richard Bentley, through private letters and conversation. His most public pronouncement of belief occurred in the General Scholium to Book III of the *Principia*. Here Newton wrote of God as follows:

The word God usually signifies *Lord*; but evry lord is not a God. It is the dominion of a spiritual being which constitutes a God: a true, supreme, or imaginary dominion makes a true, supreme, or imaginary God. And from his true dominion it follows that the true God is a living, intelligent, and powerful Being; and, from his other perfections, that he is supreme, or most perfect. He is eternal and infinite, omnipotent and omniscient; that is, his duration reaches from eternity to eternity; his presence from infinity to infinity; he governs all things, and knows all things that are or can be done. He is not eternity and infinity, but eternal and infinite; he is not duration or space, but he endures and is present. He endures forever, and is everywhere present; and by existing always and everywhere, he constitutes duration and space . . . . He is omnipresent not *virtually* only, but also *substantially*; for virtue cannot subsist without substance. In him are all things contained and moved; yet neither affects the other: God suffers nothing from the motion of bodies; bodies find no resistance from the omnipresence of God. It is allowed by all that the Supreme God exists necessarily; and by the same necessity he exists *always* and *everywhere* . . . . We know him only by his most wise and excellent contrivances of things, and final causes; we admire him for his perfections; but we reverence and adore him on account of his dominion: for we adore him as his servants; and a god without dominion, providence, and final causes, is nothing else but Fate and Nature. Blind metaphysical necessity, which is certainly the same always and everywhere, could produce no variety of things . . . . And thus much concerning God; to discourse of whom from the appearances of things, does certainly belong to Natural Philosophy.[27]

The way in which God acted in the world continued to be a source of interest and concern for Newton. The variety of activities in the world, indicated by the increasing numbers of experiments in the seventeenth century, made description of the way in which God acted difficult. There was no question of the fact of God's activity. Newton was satisfied with

God as the final cause. The process of efficient causation, however, remained unclear. Newton's suggestions for this aspect of natural philosophy all involved what he called "forces" and "spirits." The last paragraph of the General Scholium provided a good summary of the range of Newton's thoughts on "forces" and "spirits":

> And now we might add something concerning a certain most subtle spirit which pervades and lies hidden in all gross bodies; by the force and action of which spirit the particles of bodies attract one another at near distances, and cohere, if contiguous; and electric bodies operate to greater distances, as well as repelling as attracting the neighboring corpuscles; and light is emitted, reflected, refracted, inflected, and heats bodies; and all sensation is excited, and the members of animal bodies move at the command of the will, namely, by the vibrations of this spirit, mutually propagated along the solid filaments of the nerves, from the outward organs of sense to the brain, and from the brain into the muscles. But these are things that cannot be explained in few words, nor are we furnished with that sufficiency of experiments which is required to an accurate determination and demonstration of the laws by which this electric and elastic spirit operates.[28]

Newton was more specific on the source of the active spirit in nature in the text of Book III of the *Principia*. The phenomena of comets and their changing tails provided Newton with a possible source for at least one type of active spirit whose existence was indicated so clearly, he thought, by the results of observed activity on earth. While the electric and elastic spirits required further experiments, there were other observed phenomena such as vegetation and putrefaction, which indicated the presence of an active spirit operative in the air. This spirit might be supplied from the comet's tails as follows:

> ... And it is not unlikely but that the vapor (comprising the tail of the comet), thus continually rarefied and dilated, may be at last dissipated and scattered through the whole heavens, and by little and little be attracted towards the planets by its gravity, and mixed with their atmosphere; for as the seas are absolutely necessary to the constitution of our earth, that from them, the sun, by its heat, may exhale a sufficient quantity of vapors, which, being gathered together into clouds, may drop in rain, for watering of the earth, and for the production and nourishment of vegetables; or being condensed with cold on the tops of mountains (as some philosophers with reason judge), may run down in springs and rivers; so for the conservation of the seas, and fluids of the planets, comets seem to be required, that, from their exhalations and vapors condensed, the wastes of the planetary fluids spent upon vegetation and putrefaction, and converted into dry earth, may be continually supplied and made up; for all vegetables entirely derive their growths from fluids, and afterwards, in great measure, are turned into dry

earth by putrefaction; and a sort of slime is always found to settle at the bottom of putrefied fluids; and hence it is that the bulk of the solid earth is continually increased; and the fluids, if they are not supplied from without, must be in a continual decrease, and quite fail at last. I suspect, moreover, that it is chiefly from the comets that spirit comes, which is indeed the smallest but the most subtle and useful part of our air, and so much required to sustain the life of all things with us.[29]

These spirits are not limited to comets' tails:

The vapors which arise from the sun, the fixed stars, and the tails of the comets, may meet at last with, and fall into, the atmospheres of the planets by their gravity, and there be condensed and turned into water and humid spirits; and from thence, by a slow heat, pass gradually into the form of salts, and sulphurs, and tinctures, and mud, and clay, and sand, and stones, and coral, and other terrestrial substances.[30]

Newton's references to various spirits and forces were all motivated by his desire to identify and describe those aspects of nature which were other than mechanical and innate in matter. Because of his commitment to certainty in natural philosophy, and his aversion to what he considered to be hypothetical statements, he offered only those references to spirits and forces which he believed to be derived from experience or from the phenomena. He was careful to identify, and to label as such, those areas in natural philosophy pertaining to forces and spirits which were in need of further experimental investigation. The most extensive illustrations of this caution were the thirty-one queries added to his *Opticks*, as we shall see.

Newton's concern for instances of the activity of incorporeal spirits was more than a scattered inquiry into aspects of the world. Because of his view of history and his belief in God's activity in the world, Newton believed that man could obtain true knowledge, revealed to man by God through nature. Newton believed that the ancients had once possessed this true knowledge, but that it had been corrupted and lost. This knowledge included a knowledge of the active principles, forces, and spirits in nature. These were all related, finally, to God, but they often served an intermediary function, as if to free God from the necessity of acting directly, Himself, in the world, for each variation of the phenomena of nature.

God's direct activity in the world was, however, occasionally revealed. Newton believed that he had divined an example of God's action in his work on the force of gravity, and that he had rediscovered an example of God's activity which had been known to the ancients. He was sure that gravity was a universal force, and he argued for this belief in a manuscript draft of Scholia to the Propositions in Book III of the *Principia*, which included references to the ancients.

"The central purpose of the 'classical' scholia was to support the doctrine of universal gravitation as developed in these Propositions, and to enquire into its nature as a cosmic force. This doctrine is shown by Newton to be identifiable in the writings of the ancients . . . . he is not using this historical evidence in a random fashion, or merely for literary ornamentation. Rather, the evidence is used in a serious and systematic fashion, as support for, and justification of, the components of Newton's theory of matter, space and gravitation. The evidence was used to establish four basic theses, . . . These are, that there was an ancient knowledge of the truth of the following four principles: that matter is atomic in structure and moves by gravity through void space; that gravitational force acts universally; that gravity diminishes in the ratio of the inverse square of the distance between bodies; and that the true cause of gravity is the direct action of God."[31]

A careful and scholarly reinterpretation of Newton's manuscript materials by J.E. McGuire and P.M. Rattansi has served to place Isaac Newton's natural philosophy squarely amidst some important intellectual traditions of his day. They have argued convincingly that Isaac Newton should not be viewed as a "scientist," but as a "Philosopher of Nature." Newton was convinced that there was a true knowledge of the world which had been lost. It was his purpose to discover and restore this ancient knowledge, using his own powerful methods of experiment, inductive analysis, and mathematical demonstration. The purpose of his natural philosophy was most fundamentally theological. It is not possible to understand Newton's science outside of the context of his theological and philosophical thought.[32] Without the proper context, Newton the man becomes Newton the caricature scientist, with his "prism and silent face." Newton's science was only part of the range of his intellectual concerns. He was interested in alchemy, Biblical exegesis, Neo-Platonic philosophy, the early Church Fathers, the Jewish origin of Greek philosophy, and, primarily, in the restoration of an "entire and genuine philosophy" which had been lost.

Newton was in agreement with the Cambridge Platonists, especially Henry More and Ralph Cudworth, that there was a *prisca* in theology and philosophy which could be recovered. But, unlike the Cambridge Platonists, Newton added a larger element of natural philosophy to this *prisca* tradition. He believed that he could apply his method of inductive analysis to the phenomena as well as to theological and historical knowledge. In order to include natural philosophy in this "entire and genuine philosophy," Newton thought to modify the current mechanical philosophy of his day to include a concentration upon the forces of nature.

For Newton, the mechanical philosophy included matter, motion, and forces. "In one sense he expanded it, by allowing unexplained forces into his explanations of the phenomena; but in a deeper sense he restricted it, especially in its pretensions to knowledge of the natural world. A sign of this restrictive approach appeared in his early work in optics. There, he rejected the arbitrarily formulated hypotheses of such philosophers as Descartes and Hooke; for they could not from these deduce the phenomena. For Newton, the source of their error was that they did not sufficiently appreciate that the mechanical philosophy, rigorously conceived, was simply the estimation of forces in nature by geometrical calculation in terms of matter in motion. This conception was secured by the brilliant achievements of the *Principia*."[33]

Newton placed great emphasis on the study of the forces of nature. In the preface to the *Principia* he wrote that "all the difficulty in philosophy seems to consist in this—from the phenomena of motions to investigate the forces of nature, and then from these forces to demonstrate the other phenomena." But he recognized that he could only speak quantitatively about those forces which he believed could be observed from the phenomena. These were not the forces between the least parts of bodies, but were the forces which could be induced from observation of the motion of macroscopic bodies. Newton hoped to be able to extend his "estimation of forces in nature by geometrical calculation in terms of matter in motion" to microscopic bodies as well. But he realized that he had insufficient observations and experiments to enable him to do this. Forces of nature could be described, frequently with unprecedented accuracy as in the case of gravity, but Newton recognized that the nature of force was outside the range of his "experimental philosophy."

But the heart of Newton's philosophy of nature, the world of forces and active principles, lay categorically beyond the systems of the *Opticks* and *Principia*. How these principles were to be explained was a great, though hidden, problem of Newton's work. There is evidence that he tried different approaches to it at different periods, and the material of the 'classical' Scholia comes from a time when he seems to have largely abandoned earlier attempts at a quasi-material explanation of forces, and of gravity in particular. However, even when in his later years he again entertained the possibility of an 'aetherial medium,' this did not obviate the 'necessity of conceiving and recruiting it [motion] by active principles, such as are the cause of gravity. . . '.[34]

The basic problem with forces in Newton's own natural philosophy was that they could not be explained away by material mechanisms. Forces and active principles were immaterial, and therefore required the existence of an incorporeal Being for their existence and direction. The forces of nature required a different category of existence for their explanation than just matter and motion. This ontological problem of causation was central to Newton's work in natural philosophy. Knowledge of the forces was necessary to broaden the mechanical philosophy into a natural philosophy which would lead to the "restoration of the knowledge of the complete system of the cosmos, including God as the creator and as the ever-present agent."[35]

The inclusion of immaterial active principles in natural philosophy was important to Newton because, without these forces, the world would consist of matter and motion independent of God's continued intervention. To speak of forces "innate" in matter and to maintain a conception of the world as a self-operating machine was a sure road to atheism. This road was personally abhorrent to Newton, as well as being one of the widespread fears of the late seventeenth century English intellectual community.

Newton's younger disciples in the eighteenth century did not share Newton's belief in the *prisca* tradition, nor his fear of atheism, nor his view of the necessity of immaterial forces active in the world. Changing times, changing political conditions, and changing religious concerns made it difficult—almost impossible—for the young men of the early eighteenth century to understand the implications and foundations of Newton's own scientific work. What they took from the master was what they thought Newton meant, or at least what they believed he should have meant, in his statements about the physical world, as viewed from their new perspective. The Newtonians' problems of interpreting the works of the master were compounded of many ingredients. Newton was exposed to many of the tensions and conflicts between old traditions and new methods which generally characterized the Age of the Baroque. The new methods and the implications of the new natural philosophy led into areas where Newton, like Galileo and Harvey before him, was simply not disposed to follow. The "gedanken battle," waged over the comparison between the "ancients and the moderns" in the late seventeenth century resulted by the beginning of the eighteenth century in the award of victory to the moderns. This victory was declared largely on the basis of the supremacy of modern natural philosophy over that of the ancients. It is ironical that one of the exalted and most frequently evidenced examples of the supremacy of the moderns was Sir Isaac Newton and his great works, the *Opticks* and the *Principia*. Not only was Newton himself a part of the Renaissance *prisca* tradition, but his two great works were only partly indicative of his natural philosophy.

Newton consciously attempted to separate his considerations about the world into two parts—the visible, empirical, mathematically describable world and the aspects of the world which were connected more explicitly to God's worldly intervention. The world, as described in the *Opticks* and the *Principia,* seemed to be a world based on relative motions in space and time. But always in the background were Newton's basic assumptions about an absolute reference, absolute space and time, which he seemed to envision as God's sensorium. All of Newton's work was predicated upon the belief that God acted in the world, not haphazardly, but in accordance with mathematically describable "universal laws." These laws are discernible, and were once revealed to man, but man's knowledge of them was corrupted. For Newton, of course, this did not mean that God could not perform miracles. God could, and did; but He performed them for reasons of grace, not nature. God was the true cause of the activity of the world, but His ways were still imperfectly perceived. Newton consistently denied that his experimental philosophy yielded knowledge of the causes of gravity, cohesion, electricity, magnetism, and chemical activity. But he demonstrated conclusively in his *Opticks* and *Principia* that we could certainly learn more about these activities through observation and experiment. The effects could be studied and accurately described, but the causes remained outside the scope of experimental philosophy. The truth of the world was for Newton a religious truth, revealed by God. The purpose of Newton's "experimental philosophy" was to discover and to describe mathematically what could be learned from the "phenomena." The purpose of "natural philosophy," which included "experimental philosophy" as an important component, was to include God, "to discourse of whom from the appearance of things does certainly belong to natural philosophy." But ultimately, truth was revealed truth, in perfect harmony with all sources of God's revelations to man.

This theological orientation did not prevent Newton from pursuing the experimental philosophy. Quite the contrary, it provided the best of personal reasons for him to bring to bear his considerable experimental and mathematical talents. In this context, the famous thirty-one Queries added to the *Opticks* took on more than scientific meaning. They constituted serious suggestions for possible experimental paths to reveal the deeper workings of forces and active principles in nature. The study of the phenomena of light, for Newton, held the possibility of probing into the structure of gross bodies and into the forces associated with matter. These suggestions in the Queries for an experimental program provided the stimuli for a tremendous amount of experimental work in the eighteenth century. This work did result in information and description of the forces which produced observable effects. What was quickly neglected in the eighteenth century was that the experimental philosophy yielded no information about the causes of the phenomena. In fact, the causes of the forces were consciously and effectively placed outside the range of scientific activity. Eighteenth century Newtonians turned away from precisely those concerns which Newton himself considered most basic. What emerged, in the form of "modern science," was a superstructure of knowledge about effects, with the causes overtly neglected and covertly contained in hidden assumptions. Newton believed that the world was inexplicable without God. The eighteenth century Newtonians took the world as given, as described so accurately in Newton's *Principia* and *Opticks*. The problem of God's involvement lost its immediacy and was finally relegated to a realm outside that which was neatly packaged as "human knowledge."

Newton's personal theism was untenable in the eighteenth century for at least two reasons: it held the seeds of serious, even slightly ridiculous, theological difficulties, and it stood squarely in the way of the extension of Newton's methods and achievements into what we now call modern science. The image of God as the conserver of the mathematically describable order of the universe seemed to Newton and his close friends, such as Richard Bentley, an important new and true conception of God, distinct from Scholastic theology. But Newton's theism was actually a transitional stage in western theological thought, between the miraculous providentialism of earlier religious philosophy and the later attempt to identify God with the rational order and harmony of the universe. Newton's God was still providential, but one of His major duties was to maintain the exact mathematical regularity of the world. Without God's providence, the intelligibility and beauty of the world would disappear.[36] After Newton, God came quickly to assume the role of the watchmaker and cosmic tinkerer. He created the world and he continued to see that it continued to run properly. God stood continually alert to correct minor irregularities and ensure the continued perfection of His creation. During the late seventeenth century, this image was hailed as evidence of God's existence in the battle against atheism. Roger Cotes wrote in his Preface to the second edition of the *Principia*: "Newton's distinguished work will be the safest protection against the attacks of atheists, and nowhere more surely than from this quiver can one draw forth missles against the band of godless men."[37] But as the eighteenth century drew on, mechanical explanations were devised, on the basis of experiment and observation, which eliminated the need for God's attention in the world. The success of the new science in explaining the world neatly removed God from the universe.

Newton's theism also stood in the way of the new science itself. The new science forged ahead, taking Newton's experiments and mathematical demonstrations as models, and following, along with others, the paths indicated by Newton in the Queries. There was little time and less interest in questions of ultimate causation when there was literally a new world to investigate, using the new and powerful methods of science. Newton's theological reservations and serious epistemological problems were swept aside in the new enthusiasm to "get on" with science.[38] This enthusiasm over science was nowhere more evident than in the field of physical optics.

Newton left an unusual legacy to his disciples in optics. The *Opticks* itself was styled as a work which avoided hypotheses. Newton began Book I, Part I, with the following claim: "My Design in this Book is not to explain the Properties of Light by Hypotheses, but to propose and prove them by Reason and Experiments: . . ." The work consisted largely of an extensive collection of experiments with carefully compiled data. But it also contained some grand assumptions and sweeping hypotheses, Newton's protestations to the contrary notwithstanding. Most of Newton's speculations in the *Opticks* were labeled as such, however, and included in the Queries added at the end of Book III, Part I.[39] The reader was cautioned in the following manner:

> When I made the foregoing Observations, I design'd to repeat most of them with more care and exactness, and to make some new ones for determining the manner how the Rays of Light are bent in their passage by Bodies, for making the Fringes of Colours with the dark lines between them. But I was then interrupted, and cannot now think of taking these things into farther Consideration. And since I have not finish'd this part of my Design, I shall conclude with proposing only some Queries, in order to a farther search to be made by others.[40]

The area of this "farther search" was the most basic and apparently most difficult area of all: the interaction between light and matter. Newton had implied by the terms in his title that reflection, refraction, and inflection (our term, diffraction, comes from Grimaldi) were all the same type of phenomena in that they involved the interaction between light and matter in the same general way. But it was clear by the end of Book III that inflection was a special problem requiring more than the relatively simple treatment accorded reflection and refraction. The Queries contained speculations on the corporeality of light, in support of the implications in the text of the *Opticks*. But they also contained speculations on an ether, and the interaction between light and the ether, including the possible use of the ether to explain the phenomena of the colors of thin films. However, these Queries, no matter how affirmatively worded, were appended, with disclaimer, to a work which was descriptive and experimental in nature. These speculations offered no doctrinaire commitment to either the corpuscular nature of light or the existence and effects of an ether. While it seemed that Newton considered light to be corpuscular, he never offered an unqualified assertion.

The *Opticks* was so very influential largely because of Newton's reluctance to speculate in the text. It was a work which could be accepted, by those not disposed toward reading between the lines, as a marvelous example of "the experimental philosophy," with its surprisingly accurate experiments and descriptions. But, of equal importance, it was regarded by others as a mine of suggestions and speculations. Because it did not provide, by itself, a complete system of optics which offered both description and physical explanation, Newton's followers were left free to speculate as they chose on the explanations for optical phenomena. Newton provided enough hints, suggestions, and implications to make it possible for his followers to construct a corpuscular system of optics of their own devising, erected on the foundation of Newton's experiments and descriptions. Newtonian optics emerged from a patchwork of Newton's own works, held together by extensions of carefully selected threads of Newton's hints and speculations.

The founders of Newtonian optics began with the confidence that they were continuing the work of the master, based upon his own works and his own suggestions. By the 1730s, a recognizable way of viewing optical phenomena began to emerge. There were three basic features of this emerging system: (1) the consideration of light as a very small, subtle body, (2) the belief that forces of attraction and repulsion were operative between light and matter, and (3) the overall concept that the study of optics could benefit from the application of the principles of mechanics to optical phenomena. These principles of mechanics should be derived from the *Principia*, so that an explicit connection could be established between both of Newton's great works. These three assumptions became characteristics of Newtonian optics which remained both operative and recognizable until the 1830s in England. The basics of the system were formulated during Isaac Newton's lifetime.

Newtonian optics emerged within the context of the enthusiasm for atomism in England and brought the important addition of an interest in forces to this atomist tradition. By 1700, atomism was the well-accepted background for natural philosophy in England. It had ceased to be the "radical," new philosophy and had become an accepted view. The work of the Royal Society, and in particular the experimental work of Robert Boyle, had served to show that the major tenets of ancient atomism, expunged of its atheistical tendencies, were both reasonable and probable. Few could doubt the existence of atoms and corpuscles by 1700.[41]

But the existence of atoms and corpuscles had to be reconciled with the new experiments indicating the operation of forces and the probability of some ethereal medium. The Torricellian vacuum experiments and the use of the new vacuum pump raised the knotty problem of what was left when the air was removed? Experiments with static electricity, with magnetism, and with thermometers placed in a partial vacuum indicated the strong possibility of some type of ethereal medium, consisting either of individual parts surrounded by void space, or of a continuous medium. But by 1700, the Cartesian plenum was moving into the background, and the speculations on the "ethereal medium" generally assumed it to be of the form of individual particles in void space. This form was easily incorporable into atomist thought. The nature of the forces which were observed to operate remained the problem.

The Aristotelian disposition toward action by contact was difficult to abandon in the seventeenth century. This disposition toward the explanation of forces took the more detailed form of action by "impulsion" or impact of particles. This was true for the complex vortex theory of Descartes, the concepts of centrifugal forces proposed by Huygens, the notions of mechanical contacts conceived of by Hobbes, and the attempts by Boyle to explain attraction by impulsion. Regardless of the scheme devised in the seventeenth century, the corpuscularian natural philosophers were convinced that action must be transmitted by contact, and could not take place at a distance. Action-at-a-distance was rejected by everyone, up to and including Newton, as unsupportable. Action-at-a-distance conjured up the old images of sympathies and antipathies, the actions of the three kinds of souls, and various other organic analogies steeped in Aristotelian-Scholastic tradition.[42]

The problem of forces was a real and vexing one for the mechanical philosophers, and especially for Isaac Newton. How could force be explained in a way which would agree with the mechanical philosophy, involving the possibility of contact, and yet remain outside of strict mechanism sufficiently to allow for God's activity and incorporeal activity? The steps in Newton's own changing response to this problem have not yet been made explicit, but the broad outlines seem clear. Newton's early responses to the problems of forces were prompted by his proximity both to the ancient atomist tradition and to the transition from the Aristotelian-Scholastic viewpoint which was very much a part of his education. Toward the end of his life, increasing experimental evidence on electrical phenomena,[43] problems posed by inflection phenomena in optics, and the increasing interest in the physiology of vision all combined to increase Newton's interest in the ether and explanations incorporating an ethereal medium. But, whatever the changes in Newton's conceptions, one reference point remained fixed; matter alone could not account for action in the world by impulse or contact, without the activity of incorporeal spirits or principles. This, as we have seen, was essential to Newton's religion and, therefore, to his natural philosophy.

At this point, however, Newton remained on the horns of a dilemma which was largely of his own making. His careful attention to the many examples of gravitationl forces and to optics clearly indicated that most of the observed phenomena could be described, with extreme accuracy, on the basis of the action of forces of attraction and repulsion. Regardless of the form of their ultimate explanation, close observation of the behavior of gross matter yielded the undeniable conclusion that these bodies behaved as if there were attractions and repulsions between them. This was certainly true of the phenomena of gravity, electricity, magnetism, the refraction and reflection of light, and instances of chemical affinity. The description of the action of forces of attraction and repulsion, Newton believed, could be derived from experience and from the most careful observation of the phenomena.

Here Newton's experimental philosophy came into most apparent conflict with his natural philosophy. The world could be *described* in terms of matter, the void, and the forces operative between matter. But Newton knew that the world could not be so *explained.* He stated his clear reservations in this regard in the 31st Querie to the *Opticks:*

Seeing therefore the variety of Motion which we find in the World is always decreasing, there is a necessity of conserving and recruiting it by active Principles, such as are the cause of Gravity, by which Planets and Comets keep their Motions in their Orbs, and Bodies acquire great Motion in falling; and the cause of Fermentation, by which the Heart and Blood of Animals are kept in perpetual Motion and Heat; the inward Parts of the Earth are constantly warm'd, and in some places grow very hot; Bodies burn and shine, Mountains take fire, the Caverns of the Earth are blown up, and the Sun continues violently hot and lucid, and warms all things by his Light. For we meet with very little Motion in the

World, besides what is owing to these active Principles. And if it were not for these Principles, the Bodies of the Earth, Planets, Comets, Sun, and all things in them, would grow cold and freeze, and become inactive Masses; and all Petrefaction, Generation, Vegetation and Life would cease, and the Planets and Comets would not remain in their Orbs.

The men who developed Newtonian optics did not feel the same need to connect ultimate explanations about the world to the activity of incorporeal entities and God. They were enthralled by, and satisfied with, the description of the world found in the texts of the *Principia* and *Opticks*. In this sense, Newton supplied to his successors the most direct reason for disregarding what he considered to be the most important questions. He supplied them with descriptions of phenomena which worked with unprecedented accuracy. The questions which were so pressing to Newton and his generation were no longer so fraught with theological immediacy to his younger successors.

Newton's young successors had another reason to accept his descriptions of the world and neglect his explanations. They had a mathematical model of the physical world to accept in the place of deeper attempts at metaphysical explanation. This acceptance of mathematical description, to the neglect of the assumptions which lay hidden, was to be at once the strength and the weakness of modern science. It had the effect of allowing science to be done, of directing attention to those questions which could be answered within the framework of mathematical description. The amazing success of modern science attests to this strength. But the lack of personal meaning in much of modern philosophy, especially positivism and analytical philosophy, demonstrates the dangers implicit in the neglect and rejection of metaphysical concerns.[44]

The seventeenth century was astonishing in the sense that it contained two major modifications of western thought. Mary Hesse has described this transformation as follows:

... Not one, but two revolutions had taken place, both comparable in importance with the rise of Greek philosophy itself from primitive myths. Not only had every type of explanation known to the Greeks, with the exception of atomism, been entirely discredited, but the new orthodoxy, an alliance of Cartesianism and atomism, was already in decline. Newton had invented a new analogue in his mathematical theory of central forces, and a new analogue is rarer in the history of science than new theories or new methods. By the end of the century a particular type of subject-matter and a particular method of investigation were taking charge of science, and critics, though vocal, were ineffective.[45]

The "subject matter and a method of investigation" were all closely related to the mathematical model of central forces. Newton had provided a mathematical description of gravitational force which encompassed not only the motion of the planets and "Kepler's three laws," but also the motion of the moon, the tides, and the motions of comets. It was an astounding example of success, based not on physical explanations but mathematical descriptions. Newton's success with gravity and his descriptions of optical phenomena provided a choice. Natural philosophers could move away from attempts at corpuscular explanations for phenomena toward the acceptance of the mathematical description as a sufficient explanation in its own right. The new mathematical model seemed to have all the advantages. It was clear, concise, accurate, and it could be tested by observation and experiment. Because it emerged from the context of explanation in terms of matter and motion, the mathematical descriptions were assumed to have corpuscular explanations. But the major thrust of the new mathematical model was not backward into the seventeenth century to search for the details of a corpuscular explanation—it was forward into the eighteenth century, where attempts were made to extend this model to the areas of optics, electricity, magnetism, heat, and chemistry. The model served to provide accurate description, empirically observable results, and predictions for further inquiries which were themselves mathematically describable. Along with this new model went assumptions about forces, the corpuscular nature of matter and light, and the implication of straight line, action-at-a-distance. A whole corpus of metaphysical assumptions, unattended and largely unquestioned, was carried along with the success of the new mathematical model. Metaphysical questions were dismissed either as bothersome or at worst, by the end of the century, as unanswerable and therefore outside the range of concern.[46]

The Beginnings of Newtonian Optics

Specific examples of the use of the mathematical model and the acceptance of the assumptions which it implied can be found in the works of Roger Cotes and his cousin, Robert Smith. The distinction between Newton's own works and intentions, and those of these formulators of a Newtonian science become clear when we consider the beginnings of Newtonian optics. Robert Smith provided his contemporaries with an explicitly stated corpuscular system of optics. He developed a system of optics based on the concepts of attraction and repulsion between light corpuscles and matter in his textbook *A Compleat System of Opticks* (1738).[47] Smith constructed his system by making careful selections from Newton's optical works; Newton's own words and descriptions of experiments were used whenever possible. However, the context in which they were used was changed by Smith, primarily because of his attempt to draw a coherent relationship between the *Principia* and the *Opticks,* as we shall see. Smith's *Compleat System of Opticks* is considered here, not only as an important text in Newtonian optics, but also as a work which is fundamental to the understanding of the challenge and development of the undulatory theory of light by Thomas Young in the opening years of the nineteenth century. This challenge will be considered later, in Chapter III.

Smith derived his corpuscular theory by accepting as true Newton's suggestions of the corpuscular nature of light in the *Opticks*[48] and the second edition of the *Principia*. After satisfying himself that light was corpuscular from his reading of the *Opticks*, Smith proceeded to apply the concepts of attraction found in the *Principia* to optical phenomena. He adopted, in particular, the ideas of gravitational attraction and the implications of action-at-a-distance found in Cotes' *Preface*[49] to the second edition of the *Principia*, and applied them to optics, providing, for the first time, a complete, systematic treatment of reflection, refraction, and inflection in terms of the attraction and repulsion of light corpuscles by matter. It was Smith's application to physical optics of the concepts of attraction and repulsion at a distance which indicated the character of Newtonian optics in England for the next century.

Roger Cotes was especially important in shaping Smith's concepts of attraction and the action of forces. Cotes' understanding of the concept of attraction was different from that held by Newton. This difference became clear as Cotes undertook the preparation of the second edition of the *Principia*.

Cotes took great care in editing the *Principia*. Not only did he correct Newton's mathematics, but he further paid close attention to the assumptions underlying the various theorems and proofs. Cotes and Newton corresponded during the correction.[50] As Cotes prepared the *Preface* for the new edition, a question arose concerning the nature of attraction. The letters which passed back and forth between the two men show their attitudes toward the question of attraction and hypotheses in general.

In a brief outline of his *Preface* sent to Newton for approval, Cotes was troubled by Newton's concept of attraction.

I meet with a difficulty, it lyes in these words (Et cum attractio omnies mutua sit). I am persuaded they are then true when the Attraction may properly be so called, otherwise they may be false. . . . For till this objection be cleared, I would not undertake to answer anyone who should assert the You do Hypothesim fingere, I think you seem tacitly to make this supposition that ye Attractive force resides in the Central Body.[51]

Newton chose to answer Cotes in an indirect way by writing what he considered to be a hypothesis. Referring to Cotes' letter of February 18, Newton responded:

The difficulty you mention wch lies in these words (Et cum Attractio omnis mutua sit) is removed by considering that as in Geometry the word Hypothesis is not taken to so large a sense as to include the Axiomes and Postulates, so in Experimental Philosophy it is not taken in so large a sense as to include the first Principles or Axiomes wch I call the laws of motion. These Principles are deduced from Phaenomena and made general by Induction: wch is the highest evidence that a Proposition can have in this Philosophy. And the word Hypothesis is here used by me to signify only such a proposition as is not a Phaenomenon not deduced from any Phaenomena but assumed or supposed wth out any experimental proof. Now the mutual and mutually equal attraction of bodies is a branch of the third Law of motion and how they branch is deduced from Phaenomena you may see in the end of the Corollaries of ye Laws of Motion, pg. 22. If a body attracts another body contiguous to it and is not mutually attracted by the other: the attracted body will drive the other before it and both will go away together wth an accelerated motion in infinitum, as it were by a self moving principle, contrary to ye first law of motion, whereas there is no such phaenomenon in all nature.[52]

Newton considered mutually equal attraction of bodies to be deduced from phenomena. There seemed to be an experimental basis for the laws of motion. The third law was based on experience.

Experimental philosophy proceeds only upon Phaenomena and deduces general Propositions from them only by Induction. And such is the proof of mutual attraction. And the arguements for ye impenetrability, mobility and force of all bodies and for the laws of motion are not better. And he that in experimental Philosophy would except against any of these must draw his objections from some experiment or phaenomenon and not from a mere Hypothesis, if the Induction be of any force.[53]

After this rather indirect answer to Cotes' uncertainty, Newton left him to his own devices in the writing of the *Preface*.

"If you write any further Preface," Newton wrote, "I must not see it for I find that I shall be examined about it."[54]

Cotes was uneasy about being denied Newton's further comment. He sent a draft of his *Preface* to Samuel Clarke for approval since Clarke was one of the foremost advocates and defenders of Newton's works. Clarke wrote to Cotes objecting to a passage in which Cotes seemed to indicate that gravity was essential to bodies. Cotes removed the passage and replied:

> My design in that passage was not to assert Gravity to be essential to Matter, but rather to assert that we are ignorant of the essential properties of Matter and that in respect to our Knowledge, Gravity might possible lay as fair a claim to that Title as the other Properties which I mentioned.[55]

Cotes was in the awkward position of being fully capable of understanding the excellence of Newton's work on forces and being unable to appreciate the degree of the hesitancy which Newton had over the use of the concepts of action-at-a-distance and forces innate in matter. To Cotes, as to many of Newton's successors, the actions of the master spoke louder than his words. Arguments "from the phenomena" certainly indicated that forces seemed to operate. Therefore the most sensible path to follow was one of getting on with the investigations on the basis of attraction and action-at-a-distance, regardless of whether the precise causation was known. The exact mathematics helped greatly here because the descriptions of attractions matched so nicely with the observations. Not only were there the examples of gravitational forces, but there were abundant examples of references to short-range forces, convenient assumptions upon which to base mathematical descriptions. If the assumptions worked, and led to no contradictions in experience, there seemed little reason not to use them. In short, the ontological problems should not be allowed to stand in the way of epistemological investigations. Even Newton was able to forge ahead with his use of forces because they were plausible and based upon experience; they were important to his experimental philosophy. He did not claim to know the ultimate causation of gravity, or of the short-range forces of repulsion which prevented all the matter in the world from lumping together, but they were reasonable conclusions to be drawn from experience and therefore could be used. The ultimate problems did not stand in Newton's way either, but largely because he believed that ultimately God would be revealed as the final cause.

Since Cotes was left without further guidance from Newton, he found himself led to look favorably on the possibility of action-at-a-distance. In particular, his interest in refuting the Cartesian challenge to the *Principia* in the form of the doctrine of vortices in the ether led him to pose arguments against the vortices which implied action-at-a-distance. Although Cotes did not use the phrase action-at-a-distance, it seemed to him the only physical possibility if an all-pervasive fluid were ruled out. The rejection of the plenum implied the acceptance of the void and attraction. He wrote in the *Preface*:

> Those who would have the heavens filled with fluid matter but suppose it void of any inertia, do indeed in words deny a vacuum but allow it in fact. For since a fluid matter of that kind can noways be distinguished from empty space, the dispute is now about the names and not the nature of things.[56]

The comparison between Cotes and Newton is illustrative of the change of attitude in the eighteenth century. Early opposition to the concept of attraction and the void dwindled rapidly in the generation after Newton. The "uncommon incomprehensibility became a common incomprehensibility." Cotes was quite willing to "accept the force of attraction as a real, physical, and even primary property of matter." Alexandre Koyré has described this difference between Newton and the Newtonians as follows:

> Newton, himself, as we well know, never admitted attraction as a "physical force." Time and again he said, and repeated, that it was only a "mathematical force," that it was perfectly impossible—not only for matter but even for God—to act at a distance, that is, to exert action where the agent was not present; that the attractive force, therefore—and this gives us a singular insight into the limits of the so-called Newtonian empiricism—was not to be considered as one of the essential and fundamental properties of bodies (or matter), one of these properties such as extension, mobility, impenetrability, and mass, which could neither be diminished nor increased; that it was a property to be explained; that he could not do it, and that, as he did not want to give a fanciful explanation when lacking a good theory, and as science (mathematical philosophy of nature) could perfectly well proceed without one, he preferred to give none (this is one meaning of his celebrated *Hypotheses non fingo*), and leave the question open.[57]

Newton's reservations could not stand in the face of the increasing success of the "mathematical philosophy of nature," supported by observation and experiment. There was a transition from a position of reservation concerning basic, metaphysical questions to one of confident proclamation of the laws of nature, to the general neglect of metaphysical concerns.

This transition was in keeping with one of the major thrusts of the "Scientific Revolution": the establishment of the acceptability of the incomprehensible. Galileo was only among the first of the natural philosophers to argue that nature need not be what we expected or imagined her to be. The ultimate nature of the forces of attraction could remain mysterious, but the laws of nature, derived from experience and in conformity with observables, could be discovered, placed in mathematical form and used in the process of "science." The "true causes" remained obscure and untouched by the mathematical description. While Newton believed in the relationship of God to these ultimate explanations, men in the eighteenth century, with their interest focused on science, not theology, were satisfied with the acceptance of the mathematical laws. Interest centered upon what could be learned from observation, with the assumption that at some future time the causes would either become known or would at least remain equally mysterious. The positivist movement was one later manifestation of this disregard for causes. What could not be seen or investigated was part of metaphysics and should not receive attention in science. When this attitude was incautiously maintained, it led to the untenable and slightly ridiculous assumption that science and metaphysics were both separate and separable. The primary reason that metaphysics and science appeared separable in the eighteenth century was that the science contained a host of assumptions about matter, forces, space, and time which were accepted rather than actively examined. The eighteenth century demonstrated that Newton's separation of "experimental" and "natural" philosophy was perfectly viable. You could proceed to do science without continued attention to metaphysics, as long as there was a previously formulated group of assumptions to accept. The rapid progress of any field of science to the limits of its formalism is based upon this ability to set aside those questions which are unanswerable while you get on with those which are susceptible to investigation.

Cotes' position of accepting and using the concepts which seem derived from phenomena, such as attraction, did not mean that he was attempting to launch off on his own in science and break with Sir Isaac Newton. His work was illustrative of the developing ability to concentrate on "scientific" problems only. Cotes was a well-accepted, highly regarded young mathematician. He had the support of Richard Bentley, the Master of Trinity, who made him first Plumian Professor of Astronomy and Natural Philosophy in 1706, with strong support from Isaac Newton. In fact, Cotes began the second edition of the *Principia* at Bentley's suggestion. Bentley informed Newton that he should not hesitate to load Cotes with work, because of Cotes' reverence for Newton and his debt to Bentley.[58] Nor was he unaware of theological concerns. He was confident that Newton's *Principia* was one of the surest proofs against atheism, and he was ordained in the Church of England in 1713. But Cotes was interested in taking a rather practical attitude toward natural philosophy. His orientation may best be illustrated in a passage from his *Preface* concerning gravity:

> But shall gravity be therefore called an occult cause, and thrown out of philosophy, because the cause of gravity is occult and not yet discovered? Those who affirm this, should be careful not to fall into an absurdity that may overturn the foundations of all philosophy. For causes usually proceed in a continued chain from those that are more compounded to those that are more simple; when we are arrived at the most simple cause we can go no farther. Therefore no mechanical account or explanation of the most simple cause is to be expected or given; for if it could be given, the cause were not the most simple. These most simple causes will you then call occult, and reject them? Then you must reject those that immediately depend upon them, and those which depend upon these last, till philosophy is quite cleared and disencumbered of all causes.[59]

Cotes knew that primary qualities of bodies can be assumed, but not known. His arguments against vortices had served to eliminate the ether. He was convinced of the existence of attraction from the phenomena; therefore he assigned attraction as a primary quality of bodies and assumed that the concept of attraction could be used rather than impulse, because that was the alternative to the ether. Cotes became "convinced, or confirmed in his conviction, that attraction was, as a matter of fact, a property of body, and even a primordial one. Accordingly he said so in his preface."[60]

Cotes' attitudes toward attraction found their way into Newtonian optics and became one of its fundamental positions. The person fundamentally responsible for this incorporation of ideas from the *Principia* into the optics was Robert Smith. Smith was Cotes' cousin and he resided with Cotes as his assistant at Cambridge. Smith was an undergraduate and Cotes was the Plumian Professor. Early in his years of study Smith was recognized as a competent young student and was, in fact, chosen to succeed Cotes as the second Plumian Professor in July, 1716. When Cotes undertook the second edition of the *Principia*, Smith was in his second year of work toward an M.A. Because of Smith's mathematical ability and his close association with Cotes, it is almost certain that they discussed the *Principia* as Cotes worked on the new edition. Smith must also have been well aware of Cotes' *Preface* and probably discussed it with him. When Cotes died in 1716, he left his correspondence with Newton to Smith. It was Smith's preservation of this correspondence which enabled Edelston to produce his edition in 1850.[61] Smith was also responsible for the posthumous publication of some of Cotes' mathematical and scientific works.[62]

Smith was an able student of the *Principia*, apart from his association with Cotes. His selection as second Plumian Professor confirmed his competence both in reading and understanding Newton's work. Furthermore, Smith, as master of Trinity College in 1742, played a primary role in the development of mathematical interest at Cambridge. He was the founder of the famous Smith Prizes, the two annual prizes awarded for proficiency in mathematics and natural philosophy. By natural philosophy, Smith meant the works of Newton. The prizes were awarded to stimulate interest in mathematics and Newtonian science at Cambridge.

As his starting point, Smith took the notion of attraction as a primary quality of bodies and combined this with his belief in the corpuscular nature of light. He was convinced that both attraction and the corporeality of light were demonstrated from the phenomena, regardless of whether the causes had been fully revealed. In writing his *Compleat System of Opticks,* Smith developed his own concepts of attraction and the nature of light from a combination of Cotes' arguments in the *Preface* and hints supplied by Newton in his own works. Newton's Queries at the end of the *Opticks* offered some useful suggestions. The first and last of the Queries most clearly proposed the possible interaction of light and matter, although there were numerous hints throughout the Queries. The first Query asked:

Do not Bodies act upon light at a distance, and by their action bend its Rays and is not this action (caeteris paribus) strongest at the least distance?

Query 31 began:

Have not the small Particles of Bodies certain Powers, Virtues, or Forces, by which they act at a distance, not only upon the Rays of Light for reflecting, refracting, and inflecting them, but also upon one another for producing a great part of the Phaenomena of Nature?

The combination of Cotes' arguments and Newton's hints satisfied Smith. He was convinced that matter attracted light, and light acted upon matter at a distance. Smith had none of Newton's, or even Cotes', hesitancy in accepting the concept of action-at-a-distance. Smith accepted the mutual attraction of light and matter at a distance as both an accurate description and an explanation. This acceptance was to become a characteristic of Newtonian optics.

Smith was consistent in maintaining the corporeality of light. His belief in the corpuscular nature of light came primarily from Newton's suggestions in the *Opticks*; he accepted Newton's suggestions in the Queries and in the text of the *Opticks* itself as true, and built a system of optics based on the corporeality of light. He was willing to accept the attraction of light and matter and the corpuscular nature of light as physical explanations in optics. In his *Compleat System of Opticks*, Smith began the work by writing:

Whoever has considered what a number of properties and effects of light are exactly similar to the properties and effects of bodies of a sensible bulk, will find it difficult to conceive that light is anything else but a very small and distinct particle of matter: which being incessantly thrown out from shining substances and every way dispersed by reflection from all others, do impress upon our organs of seeing that peculiar motion, which is requisite to excite in our minds the sensation of light.[63]

Further, after paraphrasing Newton's Definition I of the *Opticks*, he continued,

Rays of light may, therefore, be represented by straight lines, not Mathematical but Physical, which are described by the motion of the parts or particles of light: and the point which a ray possesses in falling upon any surface may be considered as a Physical Point . . . any parcel of these . . . considered apart from the rest, is called a pencil of rays. . . .[64]

Smith's belief in the corporeality of light was clear. He combined this belief with a concept of attraction-at-a-distance, producing a corpuscular system of optics which he considered to be both descriptive and explanatory. He was able to lend support to this corpuscular interpretation in his book by paraphrasing and quoting large sections of Newton's work. He carefully selected those passages and descriptions of experiments which indicated that light was corpuscular. By conveniently avoiding any mention of Newton's speculations on properties of light which might be considered non-corpuscular in nature, Smith was able to convey the impression that Newton himself had believed light to be definitely corpuscular. No mention was made of Newton's use of "Easy fits of reflection and refraction" to explain the colors of the so-called Newton's rings or of Newton's references to the possible interaction of light and the ether. Smith simply elaborated the concept of light as a corpuscle, providing the physical explanations in terms of corpuscles and forces which Newton had so scrupulously attempted to avoid.

This, then, was one of the major differences between Newton's optics and Newtonian optics. It was clear that Newton strongly favored the concept of light as a corpuscle, but he did not commit himself in this regard. Newton represented a major transition between discussion of light as a quality and the formulation of the science of physical optics. Koyré has analyzed Newton's position on light as follows:

> Thus the famous questio disputata—whether light is a *substance* or only an *attribute*—appears to Newton to be definitely settled: light is a substance. It may even be a body, though Newton, as we have seen—and this is something that Hooke and the others will fail to notice—does not assert this outright. He believes it, of course, but he thinks that he did not demonstrate it: "body" and "substance" are not identical concepts.[65]

Smith displayed none of Newton's hesitancies. His system of optics was purportedly based on Newton's work, but it was quite different in emphasis. Smith offered a clearly corpuscular system, carefully avoiding all statements in Newton's optical works which suggested anything but the corpuscular nature of light. He did not consider a system of optics based on an undulatory motion through an ether as a possible alternative.

Smith's use of the *Principia* in his optics was another important feature. In fact, many of the assumptions of Newtonian optics are derived from the *Principia*. Newtonian optics was explicitly allied to the mechanics of attractive and repulsive forces provided by the *Principia*. Smith used the *Principia* both as a model for scientific presentation and geometric proof and as a source of inspiration for his application of the concept of forces to the various phenomena of optics. The second book of his work, *A Mathematical Treatise,* was devoted to the geometrical proof of the whole range of geometrical optics. Lenses and mirrors were treated with rigor. Smith's treatment of the theory of aberration equaled the best of Newton's proofs in the *Principia* both in difficulty and in mathematical sophistication. For the first time, Smith established a fully developed, generalized theory of the aberration of lenses and mirrors. In 1738, observational astronomers and microscopists had the limitations and theory of their instruments defined in exact and quantitative terms based on Newtonian principles. William Herschel, years later, had praise for Smith's textbook, for his treatment of aberration, and for the work in general. Herschel began his career in astronomy with a reliance upon Smith's *Compleat System*.

Smith's most interesting use of the *Principia* appeared in Book I of the *Compleat System*. This book was meant to be a popular, qualitative treatise on the properties of light, colors, lenses, mirrors, and optics in general. It was in this book that Smith applied the laws of attraction to corpuscles of light. Section XIV, Book I of Newton's *Principia* was his obvious, specific inspiration, although the whole of Book I, "The Motion of Bodies," was of general interest. Smith took the Propositions and Theorems from this section and applied them directly to geometrical optics. He was able to make this application only because he believed that light was corpuscular in nature.

Newton's Section XIV is entitled "The Motion of very small bodies when agitated by centripetal forces tending to the several parts of any very great body."[66] Newton derived the laws of refraction and reflection using the motion of these small bodies and the force exerted upon them by any very great body. Proposition XCIV Theorem XLVIII read:

If two similar mediums be separated from each other by a space terminated on both sides by parallel planes, and a body in its passage through that space be attracted or impelled perpendicularly towards either of those mediums, and not agitated or hindered by any other force; and the attraction be everywhere the same at equal distance from either plane, taken towards the same side of the plane: I say, that the sine of incidence upon either plane will be to the sine of emergence from the other plane in a given ratio.

Newton continued,

Let now the intervals of the planes be diminished, and their number be infinitely increased, so that the action of attraction or impulse, exerted according to any assigned law, may become continual and the ratio of the sine of incidence on the first plane to the sine of emergence from the last plane be all along given, will be given them also.

Fig. 1 *Transmission and reflection of light.*

This established (see Fig. 1), Newton continued with two further interesting propositions. Proposition XCV Theorem XLIX stated:

The same things being supposed, I say that the velocity of the body before its incidence is to its velocity after emergence as the sine of emergence to the sine of incidence.

And Proposition XCVI Theorem L:

The same thing being supposed, and that the motion before incidence is swifter than afterwards: I say, that if the line of incidence be inclined continually, the body will be at last reflected, and the angle of reflection will be equal to the angle of incidence.

Newton concluded this proof:

Conceive now the intervals of the planes Aa, Bb, Cc, Dd, Ee, etc., to be infinitely diminished, and the number infinitely increased, so that the attraction or impulse, exerted according to any assigned law, may become continual, and the angle of emergence remained all along equal to the angle of incidence, will be equal to the same also at last. Q.E.D.

The conclusions drawn from these proofs by Newton and Smith were quite different. Newton himself never applied his work in the *Principia* to his optics, and he would not admit a statement concerning the nature of light.

Therefore because of the analogy there is between the propagation of the rays of light and the motion of bodies [Newton wrote], I thought it not amiss to add the following Propositions for optical use; not at all considering the nature of the rays of light, or inquiring whether they are bodies or not; but only determining the curves of the bodies which are extremely like the curves of rays.[67]

Smith had none of Newton's hesitancy. He drew an exact correspondence between the motion of bodies and the motion of light. Light was a body and the rays of light were formed by the motion of those bodies. Smith's whole system was based on this extension of Newton's suggestion that rays of light could be considered composed of small bodies, acted upon by forces.

Newton also served as the source of inspiration for Smith's theory of reflection. Here again, Newton did not commit himself to his suggestion, but Smith used it as an exact statement. One of the problems which made an explanation of geometrical optics in terms of corpuscular light difficult was the explicit interaction between light corpuscles and matter. Newton made the following suggestion in the *Opticks:*

And this problem is scarce otherwise to be solved, than by saying, that the Reflexion of a Ray is effected, not by a single point of the reflecting Body, but by some power of the Body which is evenly diffused all over its surface, and by which it acts upon the Ray without immediate contact. For that the parts of Bodies do act upon light at a distance shall be shewn hereafter.[68]

Smith adopted Newton's suggestion and presented it as the explanation for the reflection of light. Smith wrote:

And this problem is scarce otherwise to be solved than by saying that the reflection of a ray is effected not by a single point of the reflecting body, but by some power of the body which is evenly diffused all over its surface and by which it acts upon a ray without immediate contact. For that, the parts of bodies do act upon light at a distance, will appear by the following experiments.[69]

The rays of light were represented by straight lines which were described by the motion of particles of light. When these particles or bodies of light were acted upon by some evenly distributed surface power of a material body, the ray of light was either reflected, refracted, or inflected. This power of a material body was not the recognized $1/r^2$ force, but some power acting more strongly over much shorter distances. The degree to which Smith adopted ideas in the *Principia* can readily be seen by this quotation:

For as light has this property in common with all other bodies, of moving straight forwards, while its motion is not distributed by any oblique force, so when it is disturbed, we may reasonably conclude, it will follow those other laws of motion to which all other bodies are equally subject.[70]

Smith considered the disturbing force to be located uniformly outside of bodies in two zones.[71] There was an outer zone of repulsion and a closer zone of attraction which existed both just above and just below the surface of the body. Since these zones of action were extremely thin and the acting forces were very strong, reflection and refraction appeared to happen at a single point rather than over any noticeable area. (See Fig. 2.)

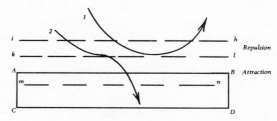

Fig. 2 Smith's explanation of the refraction and reflection of light.

Given medium *ABCD*, the process of reflection and refraction could be explained exactly. The medium had a refracting power dependent on its density; therefore, ray (1) would either be reflected or refracted depending on the "progressive power" of the ray and the force of the medium. Zone *ihkl* is the zone of repulsion, and zone *klmn* is the zone of attraction. But (1) will be the path of the ray—

if its progressive force be weak, or the repulsive force be so strong as to hinder it from entering the space of attraction *klmn*. For if it enters this space, instead of being reflected, it will be refracted into the dense medium . . . it seems to follow that the repulsive power of a dense medium is less extended or else weaker than the attractive. For if the bending of the ray by the repulsive power was not less than the contrary bending made by the attractive, the refraction into a dense medium could not always be made towards the perpendicular, as it always is.[72]

Smith offered an explanation for all known optical phenomena in terms of light corpuscles and forces. His system was complete enough to establish the applicability of a corpuscular system of optics. By use of this system, explanations could be provided for all the problems of contemporary interest. Among the most important problems in eighteenth century optics were the following:

1. Why does light travel rectilinearly?
2. How can inflection be explained?
3. How do light and matter interact?
4. How are colors produced?
5. What produces luminescence?

Smith's answers to these problems established Newtonian optics as both an explanatory and a descriptive system. He provided what has been termed a "textbook development" of the subject.[73] The textbook writer needs both a description of phenomena and a set of assumptions and explanations which make the textbook account seem both accurate and uniquely established. There is little room for hesitancies and uncertainties in a textbook. The ideas must be presented in a straightforward, determined manner. The author must decide what is true and then present it convincingly as the truth. Smith answered all of the major questions concerning optics in a straightforward manner, using the constant mode of explanation of light as a corpuscle. The problems and Smith's solutions to them will be considered in turn, to provide illustration of the Newtonian system in operation.

The rectilinear propagation of light was easily explained by Smith's corpuscular theory, using laws of motion from the *Principia*. The rays of light were straight lines because bodies tend to move in straight lines unless acted upon by an external force. Everyone who noticed beams of sunlight in a dusty room could easily observe that rays of light traveled in straight lines. The particles of light did not collide with one another when beams of light crossed because the particles were so small and the space between them so large. At that time, a simple, comparable explanation could not be given by the proponents of an undulatory theory of light.

The phenomenon of inflection had a good explanation in terms of the corpuscular theory. The corpuscles were thought to be acted upon by the inflecting bodies, such as the edges of sharp objects, at very small distances. Smith quoted several of Newton's careful experiments on the inflection of light by a sharp knife blade. Newton concluded from these experiments:

Hence I gather that the light which is least bent, and goes to the inward end of the streams, passes by the edges of the knives at greatest distance. . . . And the light which passes by the edges of the knives at distances still less and less is more and more bent. . . .[74]

Newton did not commit himself as to the cause of the inflection. He merely presented the descriptions of the experiments he performed. Smith, however, was ready to draw an explanation from Newton's experiments. "Our author had made it appear from these and some other experiments," stated Smith, "that bodies act upon light in some circumstances by an attractive and in others by a repulsive power."[75]

The experiments on inflection performed by Newton in Book III, Part I of the *Opticks* were all explained by Smith in terms of light corpuscles and forces. Newton strongly implied that light was acted upon by forces to produce inflection. Smith was more explicit; light was attracted in a specific way by matter. The power exerted by matter on light "near its surface is infinitely stronger than the power of gravity."[76] Because light was not affected by bodies as it passed as close as one-half an inch from them, but was very strongly affected by bodies when it passed extremely close to them, Smith concluded that the force, though not exactly determinable, diminished very rapidly with distance. But, at a very small distance, the force acted much more strongly than gravity. Gravity acted with a force which decreased as $1/r^2$ and yet was noticeable over great distances. The force exerted by bodies on light acted very strongly over extremely short distances, but apparently was ineffective over distances greater than an inch. Smith concluded that the force was a uniform force, similar to gravity but diminishing much more rapidly with distance.

Smith used Newton's experiments and conclusions as an example of the mutual interaction of light upon matter and matter upon light. Newton concluded that the forces of bodies to reflect and refract were nearly proportional to their densities. But, Newton concluded:

Sulphureous bodies refract more than others of the same density. . . . So since all action is mutual, sulphurs ought to act most on light. For that action between light and bodies is mutual, may appear from this consideration; that the densest bodies which refract and reflect light most strongly grow hottest in the summer, by the action of the refracted or the reflected light."[77]

Newton's experiments were considered by Smith to be experimental proof of the mutual interaction of light and matter. This is a fine example of how a prevailing assumption can serve to make the interpretation of observations and experiments subjective. With the explanation already formulated, those experiments which best matched the explanation were held up in support. This support was then considered to be conclusive evidence for the prior, assumed explanation.

Smith's use of attractive and repulsive forces in optics was one of the most explicit instances of the belief in the existence and operation of short-range forces. There had been rather widespread interest in the phenomenon of capillary action at the Royal Society in the early eighteenth century.[78] Francis Hauksbee was especially active in experiments with capillary tubes and the motion of fluids between thin plates. He used plates of glass and marble and various fluids such as different kinds of oils, water, and alcohol, and even observed the process in a vacuum to eliminate the notion that it was caused by the air. Hauksbee reported his results in papers to the Royal Society in 1709, 1711, 1712, and 1713, and in his book *Physico-mechanical Experiments*.[79] He succeeded in interesting Isaac Newton in these experiments, to the extent that Newton performed experiments of his own and included speculations on the short-range forces between matter in the 31st Query to the *Opticks* which appeared for the first time in the 1718 edition of the *Opticks*. Hauksbee wrote of Newton's work on attraction as follows:

"That very great Man, Sir Isaac Newton (the Honour of our Nation and Royal Society), has set both these Laws of Attraction in a very clear light"—namely, that amongst the greater bodies of the universe the attraction decreases reciprocally as "the Squares of the Distances do encrease", and that the smaller the portions of matter tend to each other by a law very different and unknown, but one according to which the "attractive Forces do decrease in a greater proportion than that by which the Squares of the Distances do encrease." Hauksbee then goes on to make this perfectly definite statement that "the attractive Power of small Particles of Matter acts only on such Corpuscles as are in contact with them, or removed but infinitely little Distances from them," . . . [80]

By 1720, the concept of the action of short-range forces was well accepted at the Royal Society and it was the subject of frequent discussion.

In general, it was concluded that the force of gravity was operative between large bodies according to the inverse square of the distance, and that there was another type of attractive force, operative at very small distances, to account for capillary phenomena, chemical activities, and cohesion. At extremely small distances the attractive force changes to a repulsive force.[81] Smith's treatment of optics made use of both the short-range forces of attraction and of repulsion. He extended Newton's suggestions to include the new experimental work at the Royal Society.

Smith took great pains to explain the various optical phenomena producing colors in terms of corpuscles and forces. It was in the explanation of colors that Smith relied most strongly on the corpuscular interpretation of light. The corpuscles of light composing red light were considered to be "stronger" and to possess more "progressive power" than the corpuscles composing violet.

This concept of stronger and weaker colors was clearly a throwback to the modification theory of colors put forth by Aristotle and maintained in England by Robert Boyle and Robert Hooke.[82] In the modification theory, white light was considered to be modified into the various colors by a mixture of darkness with light. Red was considered to be the strongest color because it contained the least amount of darkness, while blue, a "darker" color, was considered weaker.

Smith clearly was willing to accept the concept of strong and weak colors. He rejected the modification theory itself because Newton, as early as 1672, had argued against this theory with his new theory of colors.[83] Newton's presentation of white light had served as the reason to reject the concept of modification. But Newton had not eliminated the notions of "bigness" and "smallness," the concept of stronger and weaker colors which was a very useful conceptual scheme for the differentiation of individual colors. Newton continued to use this conceptual scheme when he treated the topic of how the eye perceives colors. The various rays strike the optic nerve and excite various vibrations, stronger or weaker depending on the color of the ray. The implication was that there was some difference in the ability of the different colors to produce these different vibrations, namely "bigness" and "smallness."

Smith developed this general line of reasoning extensively in his presentation of colors. He considered the corpuscles composing red light to be stronger, larger, and more abundant in progressive power than those composing violet. For example, the production of the spectrum of colors by a prism could easily be accounted for by considering the colors as differently sized particles of light. The red, being larger, yet having the same velocity as the other particles, would possess more progressive power than the smaller violet. The red, being stronger, would be refracted less than the violet upon passing through the zone of force.

Newton once again provided the source for this explanation. Smith quoted Newton out of context, thus attributing to Newton the concept that Smith himself presented. In both the *Opticks* and the *Compleat System* we find:

No more is requisite for producing all the variety of coulours and degrees of refragibility, than that the rays of light be bodies of different sizes; the least of which may make violet, the weakest and darkest of the colours, and be more easily diverted by refractive surfaces from its right course; and the rest are bigger and bigger, may make the stronger and more lucid colours, blue, green, yellow, and red; and be more and more difficultly diverted.[84]

But at this point the texts diverge. Newton continued:

Nothing more is requisite for putting the Rays of light into Fits of easy Reflexion and easy Transmission, than that they be small Bodies which by their attractive Powers, or some other Force, stir up Vibrations in what they act upon, which Vibrations being swifter than the Rays, overtake them successively and agitate them so as by turns to increase and decrease their Velocities, and thereby put them into those Fits.

Smith, however, supplied the text with a new conclusion:

For particles of different sizes, that fall upon the space of activity *klmn* in the line *op*, having different forces, may describe different curves, as *pa*, *pb*, *pc* and consequently will emerge from that space in different angles.

The refraction process was therefore explained by Smith in terms of the differently sized light corpuscles and the zone of force outside each refracting medium. The colors were produced by the effect of the zone of force at the surface. Smith was very careful to eliminate any of Newton's comments about "Fits of easy Reflection and easy transmission" since he wanted only to explain optics in terms of corpuscles and forces. The production of colors by simple refraction was easily explained. The process of simple reflection did not produce colors. This, too, was easily explained by Smith who argued that the space of action in the zone of repulsion of any body is so small that the components of white light, though differently affected by the repulsive force, emerge in parallel rays so close together that the appearance of a white pencil of rays is preserved.

In the refraction process, the rays of colors take diverging paths upon leaving the zone of repulsion (see Fig. 3). This divergence increases with distance, and the separation of colors is noticeable in ray (2). In the

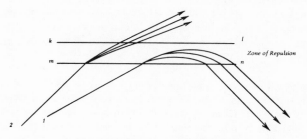

Fig. 3   Zone of force producing refraction (2) and reflection (1).

reflection process, however, the rays of colors, though separated in the zone of repulsion, are turned back in parallel paths. Smith used the analogy of cannon balls of different sizes shot from a cannon, which take different paths in flight, but strike the ground at the same angle. The rays are separated in the reflection process, but the force acting upon them is so strong, and the space of action is so small, that the separation of the parallel rays is not noticed.

The zones of repulsion and attraction were also used, as mentioned above, to easily explain the phenomena of inflection produced by a knife edge. The color fringes were explained in terms of the differing "progressive powers" of the different colors. The red, being a stronger particle with more progressive power, was inflected less than the weaker violet, accounting rather convincingly for the color fringe produced when observing inflection caused by a hair or thin wire.

To Smith, the problems of reflection, refraction, and inflection were adequately explained by action at a distance on the corpuscles of light. Of the commonly observable phenomena, only the colors produced by thin films and the combination of partial reflection and refraction at the surface of a transparent medium remained to be explained in terms of corpuscles. The phenomenon of thin films was given a very brief, but quantitative treatment. The appearance of the colored rings when two "object glasses" were compressed together was accurately described. This was the famous "Newton's rings" phenomena. Cases of both white light and monochromatic light were considered. The conclusion reached for both water bubbles and object glasses was that the colors depend on the thickness of the film or air space. This did not explain how the rings were produced, but it gave a completely satisfactory statement of the conditions and thicknesses necessary for their production. Colors were produced when the thickness was small enough. Thickness was obviously the most important consideration, made evident by the observations themselves. Smith wrote:

... the colour in every ring is the same quite round its circumference, and different rings, so it is manifest (from the spherical figure both of the object—gravity of the particles of the water subsiding gradually on all sides from the top to the bottom) that the thickness, both of the plate of air between the glasses and of the water bubble, are also the same in every part of the same ring, and different in different rings. Which shows that the particular colour of any ring depends upon a particular thickness of the plate or air or shell of water, where the incident light of open air is reflected to the eye.[85]

The emphasis placed on the measurement of the thickness needed to produce the color served as a handy and persuasive substitute for an explanation of why the colors were produced. A description was offered as a substitute for an explanation. The emphasis upon thickness agreed excellently with an emphasis on things observable from the phenomena. Here Smith had it both ways. Newton's own tentative framework of explanation involving fits and the ether was not easily reconciled with Smith's straight corpuscular approach.[86] Newton fully appreciated that an unmodified corpuscular interpretation was not completely adequate. Smith was not overly concerned with what appeared a minor explanatory difficulty. By eliminating Newton's references to "fits of easy transmission and reflection," and by dwelling upon the observations, Smith could supply an extremely accurate description which could be placed without difficulty in the corpus of description of the other light phenomena.

The question of partial reflection and transmission of light at a surface was the problem that the corpuscular interpretation could not explain. The occurrence of the colored rings could be shown to depend on thickness, but this was merely a successful evasion of the question. However, Smith noticed that "in reality some part of the incident light is always reflected and some refracted at all transparent surfaces. . . ." Realizing that he could not explain this, he continued the statement with the evasion "the cause of which our author (Newton) has also considered."[87] Smith's reference was to Newton's initial suggestion of some disposition within the ray to reflect at some times and at other times to refract.[88] This opens the whole question of the ether and the connection between the light and the ether. Newton's comments on disposition appear immediately before he launches into his discussion of the "Fits of easy transmission and reflexion." Smith was unwilling to compromise his corpuscular theory with a reference to the "Fits." But he could not avoid giving at least a reference to Newton's discussion, simply because the problems of thin films and partial reflection and transmission were serious and well-observed problems in optics. Newton recognized the seriousness of the problem by his treatment in Book III of the *Opticks*. Smith also recognized the problem, but rather than introduce an attempt at explanation which would open up the necessity of discussing the ether, he chose to keep his corpuscular interpretation intact and ignore the "Fits" in the ether. Since his uniform interpretation of optics in terms of corpuscles and forces worked so very well for all other optical phenomena, why clutter the clarity of the system because of this one problem of partial transmission and reflection? After his brief reference to the problem, Smith continued rapidly on, leaving it behind. The problem was not solved by Newtonian optics in the eighteenth century.

The more familiar phenomenon of the emission of light from bodies was easily explained by Smith, using the corpuscular theory. Connecting his conception of zones of attraction and repulsion with Newton's suggestion that heat was a form of vibration of the bodies of a material, Smith wrote:

The ray so soon as it is shaken off from a shining body, by the vibrating motion of the parts of the body, and gets beyond the reach of attraction, being driven away with exceeding great velocity.

Smith's emphasis on the corporeality of light and matter made his acceptance of the vibratory theory of heat easily understood. He answered, in the affirmative, Newton's 5th Querie on the nature of heat.[89] Smith argued that heated bodies had their parts in strong vibratory motion. Because of this motion, particles of light were shaken out of the body into the zone of repulsion and were therefore rapidly ejected as rays of light.

With only slight modification, this same type of argument was applied to luminescence. The light from Bologna phosphorus and fermenting bodies, as well as the glow produced by rubbed diamonds, was easily explained in terms of the increased agitation of those bodies which caused them to shake out light corpuscles. The corpuscles were then emitted because the action of the zone of repulsion outside the body served to drive them away.

The corpuscular theory of light proved quite adequate for both the description and explanation of the phenomena of optics in the eighteenth century. It was also compatible with other branches of physical science derived from Newton's works. Natural philosophers, especially in England, simply incorporated the corpuscular theory of light into their system of astronomy and physics. The corpuscular theory was easily maintained along with the concepts of universal attraction, action at a distance, the void, and a universe composed of hard material bodies which interacted with one another with equal and opposite effect. The corpuscular theory was adopted almost universally, while the alternate theory, involving the luminiferous ether, remained largely neglected.[90]

Newtonian optics also supported an interpretation of Newton's works termed "corpuscular dynamics"[91] which was compatible with developments in chemistry, mechanics, and physiology. Corpuscular dynamics, espoused by such men as John Keil, John Freind, and Stephen Hales, rejected the ether and even refused to include mention of the ether. Smith's *Compleat System* was very much in this tradition, except that Smith could not help including indirect references to the ether. Newton had included so many references to the ether in his optics that they were difficult to ignore. He also made it clear that there were some especially difficult problems in optics which seemed to need more than corpuscles and forces for their solution. Smith's contribution to the so-called corpuscular dynamics was his attempt to eliminate the need for the ether in optics by introducing the mechanics from the *Principia* into the field of optics. By stressing the use of forces and light corpuscles, Smith opened the way for a corpuscular system of optics which could effectively avoid subsequent references to the ether in optics.[92] Smith successfully explained, or explained away by substituting mathematical description, all of the phenomena observed during the eighteenth century. He offered, in short, exactly what his title suggested, a complete system of physical optics which combined the experiments in the *Opticks* with the straightforward explanatory scheme of corpuscular dynamics.

Robert Smith was certainly not the first to draw a corpuscular interpretation of optics from Newton's works. One of the earliest good examples of the popularization of Newton's optical works along corpuscular lines appeared in an article entitled "Light" in John Harris' *Lexicon Technicum*.[93] Harris selected those experiments and statements found in Newton's writings which made it appear that Newton had in fact said a good deal about the nature of attraction and of light. Harris presented Newton's work in Book III of the *Opticks* as follows:

He found also by plain and repeated Experiments, that the Rays of Light being in the Air, and passing near or through the Angles on any opacous or transparent Body, . . . are always *bent* or *incurvated* towards such Bodies, as if they were *attracted* by them; and of those, those Rays which pass nearest the Edges are most *incurvated*. And from hence it will follow, that the Refraction of the Rays of Light (especially those which fall near the Edges of Bodies) is not made just at the Point of Incidence, but a little before the Rays enter into the denser Medium, and a little after they are gotten within it.

The implication that matter attracted light was clear. Harris continued,

He [Newton] demonstrates also, That since Light is always propagated in Right Lines, it cannot possibly consist in Action only, (Prop. 41, 42 Lib. Princip. Phil. Mathem.) as the Cartesians do assert.

Harris clearly preferred Newton's work to that of the Continental natural philosophers, especially Descartes. Since Newton did not provide a coherent alternative to the Cartesian scheme, the popularizers were left to their own devices to present and expand carefully selected portions of Newton's writings. In order to compete successfully with the Continental schemes of explanation, Harris found it necessary to provide explanations in addition to Newton's descriptions. Having used what he could from Newton's experiments on inflection and from the two Propositions in the *Principia*, taken out of context, Harris turned to William Molyneux's work, *Dioptrica Nova*, for further support:[94] "That light is a Body, M. Molyneux in his Dioptricks proves from the various Properties of it." Harris could not find a statement by Newton which claimed outright that light was a body. The arguments which Harris outlined in support of light as a body are of interest to us because they reveal the type of thought about light current immediately after Newton's first optical papers, and the publication of the first edition of the *Principia*. Harris summarized Molyneux's arguments that light is a body as follows:

1. By the affection of its being Refracted, 'tis manifest that Light in its Passage through this and t'other diaphonous Body, does find a different Resistance . . . .

. . . And 'tis certain that *Resistance* must proceed from the *Contact* of two Bodies; and *Contact,* either *Active* or Passive, belongs *only to Body.*

The Second Property that confirms *Light* to be a *Body,* and a *Body moved* or thrust forward, is, that it requires *Time* to pass from one Place to another, and does it not in an *Instant,* but, is only of all Motions the *quickest* . . . .

A third Proof that *Light* is a *Body,* is, that it cannot by any Art or Contrivance whatsoever be *increased* or *diminished*; that is to say, we cannot magnifie (for

Instance) the *Light* of the *Sun*, or a *Candle*, no more than we can magnifie a Cubick Inch of *Gold* or make it *more* than a Cubick Inch . . . .

The whole line of argument was based upon the assumption of light as a body. The conclusions to be drawn from Harris' article are therefore rather different than those drawn from Newton's writings. Harris was forced to summarize because of the limitations of space in the *Lexicon*. He chose to emphasize what he considered to be Newton's implications about the attraction of light and matter and of light as a body.

A very similar emphasis was found in the widely read popularization of Newton's works by Henry Pemberton, *A View of Sir Isaac Newton's Philosophy*.[95] Pemberton made a very telling comment on Newton's reluctance to make statements about attraction of light and matter.

Tis true our author has not made so full a discovery of the principles by which this mutual action between light and bodies is caused; as he in relation to the power, by which the planets are kept in their courses; yet he has led us to the very entrance upon it, and pointed out the path so plainly which must be followed to reach it: that one may be bold to say, whenever mankind shall be blessed with this improvement of their knowledge, it will be derived so directly from the principles laid down by our author in this book (*Opticks*), that the greatest share of the praise due to the discovery will belong to him.[96]

It was clear to Pemberton, as to most of Newton's disciples, that the master had indicated the road of corpuscularity in his work in optics. Newton certainly would not have agreed with such an explicit statement. He believed that reason and experiment would not allow him to make final statements about the action of light and matter and about light as a body. However, the first seven Queries added to the *Opticks* indicated quite clearly to his followers that the weight of his experiments would allow conjecture along precisely those same lines.

Robert Smith was the earliest clear spokesman for the corpuscular theory. His combination of the principles of attraction applied to light with the belief that light was a corpuscle was widely received, both in England and on the Continent. After the first English edition in 1738, there appeared a German translation in 1755, two French translations in 1767, and an abridged English version in 1778. The abridged version became a popular text at Cambridge. Smith's text was also well regarded by his contemporaries. J.T. Desaguliers, the enthusiastic Newtonian, referred his readers to Smith's work on matters concerning optics in his own *A Course of Experimental Philosophy*.[97] The implication was clear that Smith's *Compleat System* was an authorized place for a student of Newtonian science to look for information on optics.

There were other works, in addition to Robert Smith's, propounding a system of corpuscular, Newtonian optics. But none of the texts in wide use after 1740 were as complete as Smith's. One of the more popular writers was Benjamin Martin, whose work *A New and Compendious System of Optics* appeared in 1740.[98] Martin's purpose was to write a textbook short enough and simple enough to present the essentials of optics to amateurs primarily interested in building and understanding microscopes and telescopes. In the interests of promoting the acceptance of his book, he offered a short—and of course biased—analysis of contemporary texts in his Preface.

If you ask why they [young men] do not study optics, they re-interrogate, what books should we read? If you refer them to Mr. Molyneaux, that is too large, and too much perplex'd with algebraical Solutions, and is therefore only fit for scholars. If you recommend Dr. Gregory's Elements, the Geometrical Demonstrations of every Proposition deter the Reader, and Mr. Brown's Supplement thereto involves him in a Labyrinth of analytical Investigations, with little Order and Perspicuity, and great want of Schemes. If, lastly, you advise them to read Dr. Smith's Treatise of Optics, they tell you it is too expensive, and so voluminous that they cannot pretend to have time for reading so much upon the Subject, besides that by far the greatest is above their Understanding.[99]

Martin's account gave a good summary of the works available to an English student of optics in the 1740s. His estimation of the various works was a bit extreme, but in general correct. Molyneux's large work offered a great deal of calculation and proof, too difficult for the reader unskilled in algebra and analysis. Gregory's work was even more mathematical. As he stated in his introduction, Gregory avoided any attempted explanation and dealt only with the strictly analytical and geometric aspects of optics "after the manner of mathematicians."

Martin's comment on Smith's *Compleat System of Opticks* was perhaps prompted by the urge to downgrade a text similar to his own. Smith's two volumes were more expensive than Martin's volume and they were far more complete. But the charge that they were written above understanding could apply only to Smith's careful analysis of the theory of aberration. The rest of the work was as clear or clearer than Martin's own work.

Benjamin Martin's works bear a brief examination here, since they are clearly representative of the Newtonian tradition of optical thought. Martin's first major publication was entitled *The Philosophical Grammar, Being a view of the Present State of Experimental Physiology or Natural Philosophy* (London, 1738). In this work, Martin was clearly committed to a Newtonian point of view. Light consisted of bodies possessing real motion upon being emitted from luminous objects.[100] These particles of light had a definite size, which could be approximated, if not exactly calculated. Light particles were considered to stream forth from a candle, for example, in tremendous numbers. His computation of the size of light particles went as follows:

A. Can you give an Account of the Smallness of the Particles of Light?
B. Yes: It is computed, that in a Second of a Minute, there flies out of a burning Candle, the following Number of Particles of Light, 41866000000000000000000000000000000000000 which is 10 000 000 000, or ten millions of millions times a bigger number than 1000000000000000000000000000000, the Number of the Grains of Sand computed to be contained in the whole Earth.[101]

Although the smallness of the particles of light could be approximated, not all the particles were of the same size, or magnitude. This was shown by a reference to the process of refraction. Martin argued that, since the rays of light were proved to be more or less reflexible and refrangible, the particles of light must therefore have greater or lesser magnitudes. The particles of red light were considered to be of the greatest magnitude and those of violet of the smallest. Martin used the terms size and magnitude interchangeably. This concept of the bigness and smallness of the light corpuscles remained a continuing part of Newtonian optics in the eighteenth century. It was a line of explicatory reasoning both persuasive and heuristically useful.

Throughout Martin's work, there was a clear assumption that matter and light interacted. He did not, however, write about the operation of this interaction as explicitly as had Smith. There were forces which were operative to produce prismatic refraction, for example, but the mechanics of this interaction were not specified. However, in 1759, Martin took a clearer position on the nature of the interaction of light and matter. His *New Elements of Optics* appeared in London in that year.[102] In this work, Martin adopted the idea of zones of attractive force above the surface of media in order to explain the refraction process. He also came to the conclusion that the principles of mechanics could be applied to the study of refraction. As a basis for the study of optics, Martin introduced ten basic propositions. Martin stated these propositions:

by way of Data, as they are Confirm'd by the most certain and obvious Experiments.

First, that the Particles of all Matter are affected with certain Powers, by which they are attracted and repelled.

... Thirdly, That this Power acts with the greatest Force upon the Surface of

Bodies, and decreases not simply as the distance increases, but according to some Power of the Distance.

... Fifthly, That a Ray of Light is real Matter, or that the Particles of Light are only Particles of Matter attenuated or subtilized to an extreme Degree.

Sixthly, That the Particles of Light and all other Matter mutually act upon each other by Attraction and Repulsion.[103]

Martin proceeded from his propositions to a consideration of Newton's knife edge experiments. He described and explained them in terms of a "space of attraction," "curves of the paths of light particles," and "accelerating forces of the medium." He concluded his discussion by saying: "All this follows from the Principles of Mechanics."

Martin's works displayed all the major characteristics of Newtonian optics by 1759. He presented his reader with the concepts of the materiality of light, forces of attraction and repulsion between light and matter, and the application of the principles of mechanics to the description and explanation of all optical phenomena. By the 1760s in England, this way of considering optics, termed Newtonian Optics, was becoming very well established and accepted. Two of the most popular textbooks, Smith's and Martin's, presented it as a well-organized system, offering both description and explanation to the student of optics. The system encompassed all known optical phenomena, and was considered to be proven by experiment. There was a great mass of accurate experimental data which was interpreted to support the notion of light as corpuscular. Those experiments which seemed to raise difficulties were played down and neglected in the literature and textbooks; they did not seem to indicate any new avenues of investigation, nor did they raise any serious enough objections to the Newtonian interpretation which could not be explained away.

Newtonian optics was made all the more secure by the support of several popular treatments of Newtonian science which were only partially concerned with optics. Such works as Richard Helsham's *A Course of Lectures in Natural Philosophy* (London, 1739), T. Rutherford's *A System of Natural Philosophy* (Cambridge, 1748) and John Rowning's *A Compendious System of Natural Philosophy* (London, 1737-43)[104] all treat light as corpuscular and within the context of mechanics and the forces operative in and between matter. A large part of the acceptability of the corpuscular interpretation of optics was its easy compatability with the assumptions of the fields of mechanics, hydrostatics, and astronomy. All of these subjects could be grouped under the same explanatory mode of bodies and forces. In these fields, the dynamic corpuscularity derived from

Newton's *Principia* and *Opticks* continued successful throughout the eighteenth century. In other fields, especially in those fields which took their inspiration from Newton's Queries, the framework of dynamic corpuscularity did not continue successfully after 1740. In particular, the fields of electricity, chemistry, and physiology adopted a materialist framework.[105] But the materialism which found its way into applications to other fields remained outside the context of Newtonian optics for several reasons. The development of optics using corpuscles and forces had been extremely successful in accounting for the vast majority of optical phenomena. There were no important embarrassing deficiencies in the system which required its extensive modification. In the face of a successful system of explanation which was compatible with mechanics and astronomy, the attempt at introducing materialism into optics in the form of a luminiferous ether was doomed to failure.[106] The ether offered no obvious advantages in optics, so there was simply no reason to change.[107] The Newtonian system of optics, whose major outlines had been established by Robert Smith in 1738, remained both dominant and effective throughout the eighteenth century.

II
The Construction of a System

Inheritors of Forces and Corpuscles

The Newtonian system of optics did not develop entirely without internal difficulties. Although the major outlines and many of the explanations were formulated by Robert Smith in 1738, there were a number of aspects of the system which needed further enquiry. Several explanations within the system were found to be open to question and some areas were found to need additional thought and experiment. The major additions and clarifications of the Newtonian system during the eighteenth century will be considered in this chapter.

One of the most important problems which arose within the Newtonian framework was the detection of a false experimental conclusion by Sir Isaac Newton himself. The problem was solved by John Dolland and led to the development of the achromatic lens. Newton had concluded that, because of the differing refrangibilities of the rays of colors, it was impossible to construct a lens free from chromatic aberration. Dolland was a supporter of this conclusion before 1752. When the famous mathematician Leonhard Euler published a paper in the *Memoirs of the Royal Academy of Berlin* in 1747 suggesting the possibility of correcting for chromatic aberration, Dolland immediately published an article in defense of Newton's experimental conclusions.[1] Dolland took the point of view that it was clearly proved by experiment that chromatic aberration was impossible to correct by any number of refractions. For his support, Dolland called upon Book I, Part 2, Prop. 3 of the *Opticks,* where Newton had written;

. . . I found myself moreover, that when light goes out of air through several contiguous mediums, as through water and glass, as often as by contrary refractione it is so corrected that it emergeth in lines parallel to those in which it was incident, continues ever after to be white. But if the emergent rays be inclined to the incident, the whiteness of the emerging light will be by degrees, in passing on from the place of emergence, become tinged in its edges with colours.

Euler had taken exception to Newton's conclusion that if light passed through several contiguous mediums, and was corrected by refraction to emerge parallel to incidence, it must necessarily be white. Within Newton's conclusion was the assumption that the various substances, such as water and air, refract light differently but disperse the colors in direct proportion to the degree of dispersion. The relative amount of dispersion of colors, one to another, was assumed by Newton always to be constant. If this assumption were true, it would be impossible ever to correct lenses for chromatic aberration by any number of refractions whatsoever.

This assumption was natural in the context of the representation of colors by corpuscles of a well-defined size. Since the ratio of the "bigness" or progressive powers of the corpuscles was uniform, it should remain constant. If the forces changed from one medium to another, the forces would act uniformly upon the corpuscles of light. The action upon the rays of light by any given medium should therefore produce the same ratio of refraction.

Euler took exception to this assumption by deriving a theorem which predicted that lenses could be made free from aberration by a combination of refractions in different media. He based his calculations on the ratios of refractions of light by water and air. Euler represented these ratios in a way which he found to be theoretically correct. They were not the ratios, however, which could be derived from Newton's experiments or from Newtonian assumptions. Euler argued that by the combination of the properties of air and water, light could be made to focus achromatically.

Dolland took exception to Euler's calculations because he asserted that Euler's mathematical assumptions and proportions were not in agreement with previous experimental results.[2] Dolland argued that Euler's theorem was based on Euler's own law of refraction, created by himself. The results of previous experimentation, particularly that of Sir Isaac Newton, would not support the theoretical argument set forth by Euler. He concluded by saying that Euler's calculation of a new theorem for lens construction was purely his own creation, not supported by experiment. Since it was not supported by previous experiment, it could not be valid. It was therefore still impossible to correct the aberration of object glasses by any number of refractions whatsoever. Euler's work had all of the appearances of a hypothesis unsupported by the phenomena. Dolland simply preferred to take Sir Isaac Newton's conclusions on faith.

Dolland's preference for what he thought to be experimental evidence led him to deny any possibility of correcting aberration in 1753. But in early 1755, his doubts were renewed by the work of a Swedish mathematician, Samuel Klingenstierna. The Swede found, as a result of geometrical investigation, that Newton's law of dispersion led to

The Construction of a System

predictions which did not match observations.[3] This was precisely the type of argument which Dolland had used against Euler. Dolland now decided to perform some experiments of his own to test the validity of the accepted law of dispersion which was derived from Newton's experiment. The results of his renewed interest in refraction were published in the *Philosophical Transactions of the Royal Society* in 1758.[4] Dolland succeeded in proving to himself that the accepted law of dispersion was incorrect. This overthrow allowed him to begin the search for a way to produce an achromatic lens.

The surprising aspect of Dolland's discovery lay in the fact that he simply repeated Newton's experiment to obtain the new result. Dolland recorded his surprise as follows:

This experiment will be readily perceived to be the same as that which Sir Isaac Newton mentions; (Book I. Part ii, Prop. 3 Experiment 8 of his *Opticks*) but how it comes to differ so very remarkably with the result, I shall not take upon me to account for.[5]

The experiment consisted of a double prism of glass and water. Two plates of glass were cemented together to form a wedge-shaped vessel, closed at both ends. A glass prism was placed within the vessel and the vacancy filled with water. The refraction of the prism was arranged to be contrary to that of the water, so that the contrary refractions were made to be equal and the emergent rays were parallel to the incident.

According to expectations, an object viewed through such an arrangement would have its natural color. Since the rays emerged parallel to their incidence, they were not supposed to produce color by their experience of refraction. Dolland found this to be definitely not the case. The assumption had been, since Newton's time, that equal refractions would yield equal dispersions of colors. Dolland found, after repeated experiment, that the dispersion of light into colors under equal refractions was certainly not the same for all substances. He discovered a new principle of dispersion: the dispersion of the incident light into colors was dependent on the refracting substance itself. The same colors of light were not treated alike by all refracting substances, as would be the case if they had their own refrangibility strictly associated with each separate color. Dolland found that the various colors had different refrangibilities according to the substance by which they were refracted. Therefore, even though the rays of light came out parallel as a result of their double refraction by water and glass, they were not unchanged as had been expected, but were rather distinctly colored. The divergence of the colors was by no means in proportion to the refractions, which were, of course, different in glass and water. The dispersion took place differently in glass than in water.

Based on the experiments he had performed in 1757, Dolland then began to grind telescope object glasses according to his new principle of refraction. According to his principle, it was clearly possible to find an arrangement by which the differing dispersions would correct themselves and the light would pass through the object glass without the aberration produced by colors. His first lenses were arrangements of two spherical glasses with water between them. But, much to his disappointment, while the rays emerged without color, they would not focus properly because of the old problem of spherical aberration, an error produced by the inaccuracy of the old law of dispersion derived from Newton's experiments. There were some men, however, who challenged the novelty was impossible to grind the lenses accurately enough.

Dolland was disappointed by the finding. Because of the inaccurate lens-grinding techniques of the time, spherical aberration presented a very definite problem. Lenses of short focal length simply could not be ground accurately enough to produce a sharp focus. The alternative was now apparent: to look for different refracting substances which would yield both the advantages of a color-free image and the possibility of longer focal length, thus allowing the lens to be ground flatter. Dolland was certain this could be accomplished because his experiments had clearly shown that different substances made light diverge differently, in proportion to their index of refraction.

His search for more suitable substances led to immediate results. He wrote of his search:

I discovered a difference, far beyond my hopes, in the refractive qualities of different kinds of glass, with respect to their divergency of colors. . . . It was not now my business to examine into the particular qualities of every kind of glass that I could come at, much less to amuse myself with conjectures about the cause, but to fix upon two sorts as their difference was the greatest; which I soon found to be crown, and the white flint or crystal.[6]

Dolland succeeded in discovering both the principle and the materials necessary for the construction of achromatic lenses. There followed considerable technical difficulty in the actual production and installation of these lenses, but these were overcome by the combined lens-making skills of Dolland and his son, Peter. By 1765, achromatic lenses of great accuracy were available to all those who could afford them.[7]

The Construction of a System

Dolland's discovery clearly proceeded from the realization of the inaccuracy of the old law of dispersion derived from Newton's experiments. There were some men, however, who challenged the novelty of Dolland's work once he had succeeded. The claim was made, on behalf of Sir Isaac Newton, that Newton's own experiments were quite sufficient to provide the correct theory of achromatic lenses. Joseph Priestley, a contemporary writer, related the situation as follows:

> Notwithstanding it evidently appeared I may say to almost all philosophers that Mr. Dolland had made a real discovery of something not comprehended in the optical principles of Sir Isaac Newton, it did not appear so to so sensible a man, and so good a mathematician as Mr. Murdock is universally acknowledged to be.[8]

Murdock presented the case for the correctness of Newton's work in a very detailed paper to the Royal Society of London.[9] Murdock deduced the proper law of dispersion by reading it into Newton's Experiment number VIII, introducing conditions and assumptions that simply were not there. His attempt appears to have been purely for the purpose of shielding Sir Isaac's good name from any slight shadow of inaccuracy. The paper was of no consequence but may be seen as an example of the extent to which Newton's self-styled disciples would go in their effort to save the master from criticism.

Sir Isaac Newton's reputation was immense and a subject of national pride in England. Frank Manuel has summarized the popular conception of Newton vividly as follows:

> By the early eighteenth century, Newtonian science had acquired many faces, and it showed them all: For the young scholars, it was a scientific philosophy; for the bishops, it was a proof of the existence of God; for the merchants, it offered the prospect of reducing losses at sea; for the King, it was an embellishment of the throne; for aristocrats, it was an amusement. Thus, it could be assimilated in many different forms—not excluding "Newtonianism for the Ladies"—a protean quality that is almost a prerequisite for universalist doctrines. In order to secure itself, the science of Isaac Newton used certain of the mechanisms of a conquering new religion or political ideology. It triumphed, a truth in its day, but it seems to have availed itself of the same apparatus as any other kind of movement. Followers were assembled and bound to an apotheosized leader with ties of great strength. An internal institutional structure was fortified. The word was propagated by chosen disciples. Since the doctrine was rooted in a national society, its relations with the government gave it special privileges and emoluments. It became the second spiritual establishment of the realm, and at least in its origins presented no threat to the primary religious establishment that it was destined to undermine and, perhaps, ultimately to replace. As in many militant doctrinal movements, the truth was not

allowed to fend for itself, and on occasion the sacred lie and the pious fraud became means to a higher end.[10]

This picture of Newton as national hero is overdrawn, but it is rather accurate when applied to those men on the fringes of scientific work who were not capable of original contribution. Isaac Newton's dominance did not extend to the "good" disciples who were capable of going beyond the works of the master. It also did not extend to those areas not covered in detail in Newton's works; to electricity, physiology, and the like. These areas were shaped, but hardly dominated by Newton's influence. The major thrust of Newton's influence came in the public conception of science in England.

Esteem for Isaac Newton found its way into literature and especially poetry. The poets were particularly fond of the *Opticks* because of its revelation of the nature of white light, its experiments with prismatic colors and its treatment of the rainbow. "Widespread interest of poets in the *Opticks* began in 1727, at the time of Newton's death, when the feeling for 'Britain's justest pride' amounted almost to deification."[11] After Newton died, many poets wrote eulogies stressing the importance of both his *Opticks* and the *Principia*. Newton's public reputation was so great that in a literary age which loved satire and which preyed upon philosophers and natural philosophers alike in satirical works, Newton remained above and beyond satire; "But the 'godlike Newton' remained somehow apart, beyond evil, beyond satire."[12] When the popular writers did venture to disagree with Newton, they did so respectfully.

The scientists were as careful as their literary colleagues in their treatment of Isaac Newton. The example of the law of dispersion seems to have been the only genuine inaccuracy recognized in Newton's works. But there were, however, several other aspects of Newtonian optics which fell under question in the eighteenth century. In particular, the whole structure explaining the process of refraction of light was called into question by one Thomas Melvil.

Melvil was dissatisfied with the belief that light particles possessed different sizes. He considered it more probable to assume that the differently colored rays of light had different velocities of emission from luminous bodies. He attempted to show that different velocity rather than different size should be used in the Newtonian system to explain the refrangibility of light.[13]

Melvil's criticism was actually a twofold criticism of the existing Newtonian system. His suggestion for the substitution of velocity for size to explain the dispersion concept in refraction was linked to his attempted

The Construction of a System

revival of consideration of Newton's concept of "Fits of easy transmission and reflexion" of light. Melvil assumed that particles of each of the colors were emitted with different velocities—red the swiftest, and violet the slowest. If the various colors did have different velocities, then the pulses in the ether, which comprised the "fits" must have different "intervals," as Melvil called them, to accommodate the arrival of the different rays at the same boundary. Melvil put the argument as follows:

> While the differently-colour'd rays are supposed to move with one common velocity, any pulse, excited in the aethereal medium, must overtake them at equal distances; and therefore the intervals of the fits of reflexion and transmission, if they arise in this manner, as Sir Isaac conjectures, would all be equal: but if the red move swiftest, the violet slowest, and the intermediate colours with intermediate velocities, it is plain, that the same pulses must overtake the violet soonest, the other colours in their order, and last of all the red; that is the intervals of the fits must be least in the violet, and gradually greater in the prismatic order, agreeably to observation.[14]

Melvil followed this reasoning with an elaborate calculation concerning the relative velocities of the colors and the "aethereal pulses." He concluded that the velocities of red and violet in air are nearly as 78 to 77. The velocities of the "aethereal pulses," this time in celestial space, came out to be to that of red light as 79,763 to 78,000.

Melvil further suggested that his ideas be put to experimental test by observation of the satellites of Jupiter. The consequence of Melvil's ideas would be a change of color, through the colors of the spectrum, as the satellite submerged behind the planet and again as it emerged on the opposite side.

> ... the last sensible violet-light which the satellite reflects before its total immersion into Jupiter's shadow, ought to continue to affect the eye for a 77th of 41'; that is, about 32" of time after the last sensible red light is gone.[15]

Melvil agreed that the time was long enough to be directly observable. He concluded the paper with the argument that if the phenomena was observed by astronomers, his suggestions were true. If it was not observed, then the rays "of all colours are emitted from the luminous body with one common velocity."

Thomas Melvil had expanded upon his suggestions at length in two papers before the Royal Philosophical Society of Edinburgh on January 3 and February 7, 1753.[16] He was a staunch supporter of Newtonian optics, particularly against such Continental opponents as Leonhard Euler, but he was concerned that refraction might be imperfectly understood and he was particularly amazed that the understanding of the phenomena of inflexion had not been better advanced since Newton's death. In an attempt to provide this advance in the understanding of inflexion, Melvil re-opened a consideration of the "fits of easy transmission and reflexion," something earlier Newtonians such as Smith and Martin had been extremely reluctant to do and had meticulously avoided. Melvil's discussion of the "fits" was clearly motivated by his desire to advance an area in optics which Sir Isaac Newton had specifically left to his followers to continue. But this discussion in 1752 was also conducted at a time when the ether was an acceptable, in fact popular, topic for discussion, in contrast to Smith's day when arguments against the Cartesian vortices were still an important part of Newtonian science. Melvil considered the "fits" as fair game for conjecture. He wrote, in proper Newtonian manner, as follows:

As it is of great consequence in philosophy to distinguish between facts and hypotheses, however plausible; it ought to be observed, that the various refrangibility, reflexibility, and inflexibility of the several colours, and their alternate dispositions at equal intervals to be reflected and transmitted, which are the whole ground-work of the *Newtonian* system, are to be considered as certain facts deduced from experiment: but whether the velocities of the different rays are exactly equal, or different in the manner now described, is no more than probable conjecture; and, tho' this point should be decided by a method proposed afterwards, it would still continue uncertain, whether the fits of reflexion and transmission are occasioned by an alternate acceleration and retardation of the motion of light, or in some other manner.* (*For instance, it might be supposed, that every particle of light has two contrary poles, like a load-stone; the one of which is attracted by the parts of bodies, and the other repelled; and that, besides their uniform rectilineal motion, the particles of differently-coloured rays revolve in different periods round their center: for thus, their friendly and unfriendly poles being alternately turned towards the surfaces of bodies, they might be alternately disposed to reflexion and transmission; and that at different intervals, in proportion to the periods of their rotation.) And, after all, it is no more than probable conjecture, that such an alternate acceleration and retardation is brought about by the influence of pulses excited in the ethereal *medium:* nay there are some circumstances in these *Phaenomena* that seem hardly intelligible by that hypothesis alone; as, why the intervals of the fits are less in denser *mediums;* and why they increase so fast and in so intricate a proportion, according to the obliquity of incidence.[17]

Newton's suggestions of the "fits of easy transmission and reflexion" in Book III of the *Opticks* were difficult to avoid if the topic of inflexion was to be discussed. Melvil wanted to consider the subject and added twenty-four Queries of his own to his paper, addressed mainly to the topic of inflexion. This was not a "heretical" thing for a Newtonian to do, but it

The Construction of a System

was not very useful either. Other Newtonians had also mentioned the "fits." Perhaps the most famous reference to the etherial medium and "Fits" occurred in Henry Pemberton's widely read *A View of Sir Isaac Newton's Philosophy*. Although this work was primarily concerned with the *Principia*, there was a sizable section assigned to the *Opticks*. In general, Pemberton gave what he considered to be a summary of the salient points of the *Opticks*, including a brief presentation of what he considered to be Newton's position on the connection between light, bodies, and the ether.

"What the power in nature is, whereby this action between light and bodies is caused, our author has not discovered," Pemberton wrote. "Sir Isaac Newton has in general hinted at his opinion concerning it; that probably it is owing some very subtle and elastic substance diffused through the universe, in which such vibrations may be excited by the rays of light, as they pass through it, that shall occasion it to operate so differently upon the light in different places as to give rise to these alternate fits of reflection and transmission."[18]

The conclusions to be drawn from Pemberton's comments were that Newton had not arrived at a satisfactory conception of the nature of attraction and that an etherial medium and "Fits" were his tentative conclusions. Pemberton considered Newton's work on the attraction of planets and large bodies both conclusive and proven. However, he did not assign a degree of finality to Newton's statements on small-scale attractions, such as those between light and bodies. Pemberton preferred to indicate Newton's thoughts on "Fits" and the ether, giving the impression that the question was by no means settled.

Melvil's criticism, then, can be seen as a continuation of the lines of thought which gave some degree of acceptance of "Fits." The association of Melvil's ideas on "Fits" with his new suggestion on the different velocities of light seems to have been an unfortunate one for the attempt to use the ether in optics, because Melvil's suggestion was soon decisively proven erroneous. The concept of "Fits" and the inclusion of any suggestion of an ethereal medium faded more and more into the background as the century progressed.

Melvil's speculation was proven false by James Short. He set out to observe the satellites of Jupiter immediately after he heard Melvil's suggestion of the color change upon eclipse. He concluded after observation:

So that, upon the whole, we may conclude, that it does not appear, by the observations of the emersions of the first satellite of Jupiter, that the rays of different colours move with different degrees of velocity.[19]

Short added that his experiments showed that the colors of rays of light have the same velocity, only if the assumption was made that light was "propagated by a continued motion, in the manner of a projectile." The different velocities could still be proposed if light were considered to consist of different velocities of vibration, in the manner of the vibrations of sound in air. The experiment was therefore a decisive one only for those persons partial to the Newtonian system of optics.

Dolland's work on the principle of dispersion settled the question with more finality. According to Joseph Priestley in his *History of Vision, Light and Colours,* Dolland's experiments

clearly prove that the different refrangibility of the rays of light depends upon properties that are independent of different velocity; since the proportion of it varies according to the nature of the substance it falls upon. The same is also manifest from the consideration, that the distances at which differently coloured rays are attracted and repelled are different.[20]

The question raised by Melvil appeared to have been decisively settled by Dolland's correction of the law of dispersion, and by Short's observations. The various sizes or magnitudes of the particles of colored light were held to be the factors governing the effect of the forces of matter upon light. The suggestion of differing velocities for the colors as well as the use of "Fits of easy transmission and reflexion" were banished from the Newtonian system. The sizes of light corpuscles and the uniform velocity of the various colors and forces acting upon them became established concepts in Newtonian optics.

The acceptance of light as a body and the belief in its great velocity raised the rather obvious question of the momentum of light. It was puzzling that the momentum of light was so difficult to detect. Here the Newtonians had to proceed cautiously. The momentum could not be too great, because of the consequent destructive capability of light striking objects, but it should at least be observable. Joseph Priestley, a proponent of the Newtonian system of optics, wrote the following:

Admitting the materiality of light, it must, however, be acknowledged that the particles of which it consists are extremely minute, and, not withstanding its amazing velocity, that its momentum is very small.[21]

... It would certainly go a great way toward proving the materiality of the rays of light if it could be observed that they had any momentum, so as, by their impulse, to give motion to light bodies.[22]

This argument had occurred to several other investigators as well, and Priestley's *History* contains the best and most thorough account of these

The Construction of a System

men's eighteenth century investigation of the momentum of light. Among the investigators were two Frenchmen, William Homberg and Dortous de Mairan. Both pursued experiments to determine the momentum of light without reaching any definite conclusions because of the great experimental difficulties involved. The two main difficulties they encountered arose from the necessity for extremely sensitive instruments. The greatest problem was the production of convection currents in the air, adversely affecting the delicate instruments. Light had to be focused upon the instrument to show momentum. The focused light produced a heating effect in the air which destroyed the experiment. The second problem was the destruction of the apparatus itself by the heat which the focused light produced. In attempting to strengthen the effect to overcome convection problems, the intensity of the focused light was increased. This resulted in great heat which eventually melted the apparatus.

William Homberg had tried experiments to detect the momentum of light in 1708. They were conducted by using a straightened watch spring with one end embedded in a block of wood. Light was focused by a large lens on the free end of the watch spring. The watch spring vibrated after the light was focused upon it and some generally favorable, qualitative conclusions were drawn as to the slight momentum of light.[23]

Dortous de Mairan attempted to refine Homberg's experiments by eliminating some of the problems of currents in the air. He conducted a number of different experiments, all of which gave some tentative indications that light possessed momentum, but all of which also encountered the same problem of air in motion which cast doubt on the conclusiveness of the experimental evidence. De Mairan's failure to obtain experimental results did not dissuade him from his original belief in the corpuscular nature of light. He had initiated his experiments, after all, to demonstrate this very assumption.

"Light is certainly a body," he wrote, "since it affects bodies, such as our organs. It thus has an impulsive force against the bodies which it encounters on its path, if it moves and it moves, since it comes from the sun to us."[24]

A group of Englishmen were equally interested in the question of the momentum of light because of this topic's importance to the assumption of the corpuscular nature of light. The question of momentum of light was seen in the broader context of dynamic corpuscularity and the actions of forces and bodies in a Newtonian universe. John Michell, Henry Cavendish, and Joseph Priestley were concerned with the maintenance of a mechanist interpretation of the fields of optics, mechanics, and astronomy.

By the mid-eighteenth century there were two rather distinct lines of development in natural philosophy, both claiming as their mainspring of inspiration the works of Sir Isaac Newton. The two lines of development were both equally plausible, but they were effective in their application to different fields of natural philosophy. Newton's wish to refrain from hypotheses and his deliberate avoidance of unequivocal statements about the causes of forces and the nature of matter, allowed two different overall views of his works, dependent upon the current interests of the interpreter. Robert Schofield has summarized the situation succinctly as follows:

> Now it is only by implication, and that not a clear one, that Newton's theory of matter can be determined. There is no doubt that he was a corpuscularian, nor that he had modified that belief, rejecting the notion that all natural phenomena were explicable simply in terms of the various sizes, shapes, and motions of these fundamental particles of nature. But Newton scholars are still divided as to whether, in the end, he believed that the corpuscles also acted upon one another, at a distance, by means of unexplained, immaterial forces of attraction and repulsion, or that an intermediary aether subtle, elastic, and electric, provided the mechanism for their action. For our purposes, the answer to this problem is essentially irrelevant for we need rather to know that eighteenth-century natural philosophers believed that Newton believed. Unfortunately, it appears that this conflict of opinion divided eighteenth- as well as twentieth-century Newtonians. In the long run the most influential view was probably that of the aetherial school, in which more-or-less traditional materialists successfully reconciled their views with Newton's aether into a series of imponderable fluids each of which carried the properties essential to explain the various phenomena they had been created to solve. Nevertheless, there remained a clear line of British investigators, starting early in the century, who adopted the notion of forces and ignored that of the aether.[25]

These two lines of development persisted until the end of the century, and it is thus plausible, as Schofield has argued persuasively, to view the work of late eighteenth century natural philosophers from the broadened perspective of their possible philosophical assumptions, in addition to their extensive experiments and observations.

From this broader perspective, it becomes apparent that the kinds of underlying assumptions made in natural philosophy in England fall into two major categories with general, though still meaningful characteristics. The characteristics which can be grouped with the category which used materialist explanations are the use of various imponderable fluids, emphasis upon an empirical approach, and a lack of mathematics. The category which ignored the ether and relied upon forces also relied more heavily upon mathematics and theoretical calculation.

The Construction of a System

The mode of explanation which used imponderable fluids gained widespread acceptance after the 1740s largely because there were popular areas of experimental investigation which found the mechanist, corpuscularian interpretation to be inadequate. This was expecially true of investigations in chemistry, electricity, magnetism, heat, and physiology. There was a move toward an explanatory mode reminiscent of the Aristotelian qualities which carried along with them the desired explanation. "The fluid of heat, for example, does not heat because it is imponderable, tenuous and possesses certain attractive and repulsive powers; it is imponderable, tenuous and possesses these particular powers because it is the fluid of heat. The heating is a function only of its addition or subtraction from a body."[26]

The mechanistic approach had failed to provide suitable schemes to meet the new demands which emerged from experiments in the first half of the eighteenth century. This was particularly true in the areas of chemistry, electricity, magnetism, and heat, where corpuscularian dynamics could not be made to work. It was in these areas that men using materialist arguments were able to achieve answers to questions which the dynamic corpuscularians could scarcely even define. Quality-bearing substances proved much easier to conceive of and use to explain observations than were quality-causing mechanisms, hindered now by their previous strong point, mathematics. The new imponderable fluids provided easy, qualitative explanations for a host of topics not capable of inclusion in the quantitative scheme of mechanisms and forces.

Mechanistic arguments did continue successfully, however, in the study of optics, mechanics, and astronomy. Indeed, all the attempts to introduce materialistic explanations into these fields failed in the eighteenth century because they were not accompanied either by a successful explanation of a previously unexplained phenomenon or by any new mathematical techniques or models. The attempted reversion to the undulations in the ether to explain optical phenomena offered no advantages and seemed to go against the weight of experimental evidence. (This situation changed dramatically, as we shall see, in the early nineteenth century.) However, after 1740, there were still some natural philosophers who resisted the move toward materialism and attempted to maintain the line of development of corpuscular dynamics.

The work of John Michell and Henry Cavendish is particularly interesting because it bears close resemblance to the work performed in optics at the beginning of the period of Newtonian optics. Both Michell and his friend Cavendish rejected the materialist trend of explanation and adhered to what they believed to be the line of natural philosophy indicated by Isaac Newton in the *Principia* and *Opticks*. Russell McCormach has provided an analysis of their philosophy as follows:

> They believed with Newton that the purpose of natural philosophy is to discover the mathematical laws of the attracting and repelling forces between the particles of bodies, and to deduce new phenomena from the forces. They stood out against the current of British natural philosophy, which by then was neither mathematical nor primarily concerned with laws of force. The best science of the day was onesidedly experimental; and the speculative interest was generally focused on the ether, and on force-denying mechanisms, and on the physical connections between the post-Newtonian weightless fluids.
>
> Michell and Cavendish attempted to extend Newton's achievement by following the philosophical path that led to it. The force of gravity being known, the remaining object of Newtonian gravitational philosophy was to deduce new phenomena from it. Michell and Cavendish followed this directive in their search for new stellar, terrestrial and optical effects of gravitation; . . .[27]

Both Michell and Cavendish have been described as the "last of the corpuscularians"[28] because of their maintenance of interest in forces, mathematical representation, and the motion of corporeal bodies. They attempted to construct a unified conception of the world, modeled after the treatment of bodies and forces in the *Principia*. Both men took Newton's suggestion of the investigation of the forces of nature to reveal truths very seriously. They were equally impressed by Newton's suggestions, especially in the Queries to the *Opticks*, that the program of investigation of forces could be extended from gravity to include electrical, magnetic, and optical phenomena. They were also interested in the action of short-range forces of attraction and repulsion, especially in the fields of optics, heat, and pneumatics. This approach to the physical world was particularly appealing to them because it lent itself to mathematical treatment. Their approach was entirely consistent with that developed by Roger Cotes and Robert Smith earlier in the century and is a fine example of the continuity achieved by the combination of Newton's *Principia* and *Opticks* into one viable approach to the understanding of nature.

Both Michell and Cavendish were believers in corpuscular optics. Their optical investigations were based upon the assumption that light was a body, acted upon by forces to produce the variety of observed phenomena. Michell's work in optics is known primarily through the favorable account of his work presented by Joseph Priestley in his *The History and Present State of Discoveries relating to Vision, Light and Colours*

(London, 1772). Priestley depended heavily upon Michell's advice and contributions in the writing of this work.[29] Michell's interest in optics covered the range of current concerns. He was familiar with the suggestion that the colors of light depend upon differences in their velocities. He strengthened the rejection of this concept by an argument from sidereal astronomy. The rays of colors could not possess different velocities because, if they did, stellar aberration should be greater for the blue light than for the red light emitted from the stars. This was not observed, so the colors must have uniform velocities.

John Michell was also interested in the experiments to determine the momentum of light. Priestley recorded that Michell attempted to improve upon the experiments of Homberg and de Mairan, and succeeded in doing so. He constructed a piece of apparatus delicate enough to indicate the momentum of light and almost free from the obvious problems of convection currents. The experiment could be performed three or four times over before the apparatus melted in the intense light, was remarkable for its delicacy, and was of unprecedented sensitivity for its time. It consisted of what amounted to a torsion balance: a very thin copper plate about one inch square fastened to a thin harpsichord wire about ten inches long.

To the middle of this wire was fixed an agate cap, such as commonly used for small mariners compasses, after the manner of which it was intended to turn; and at the other end of the wire, was a middling sized short corn, as a counterpoize to the copper plate. This instrument had also fixed to it in the middle, at right angles to the length of the wire, and in a horizontal direction, a small bit of very slender sewing needle, about one third, or perhaps half an inch long, which was made magnetical. In this state the whole instrument might weigh about ten grains. It was placed upon a very sharp pointed needle, on which the agate cap turned extremely freely; and, to prevent its being disturbed by any motion of the air, it was included in a box, the lid and front of which were of glass. This box was about twelve inches long, six or seven deep, and about as much in width; the needle standing upright in the middle.[30]

The needle was aligned by the magnet so that the copper plate was perpendicular to the sun's rays. The sunlight was focused upon the plate by means of a concave mirror about two feet in diameter. The rays of light struck the plate and caused the apparatus to rotate slowly, "about an inch in a second of time." The experiment could be repeated several times, until the copper plate warped and melted. Priestley was convinced that Michell's experiment demonstrated the momentum of light, and also the validity of the assumption of the corpuscular nature of light. Michell's work also provided Priestley with data to refute two objections to corpuscular optics which had been raised by Leonhard Euler and Benjamin Franklin.

Both Euler and Franklin were advocates of a wave concept of light. Both argued that if light were a body, the sun and the stars would continually lose mass because of the light emitted by them and therefore decrease in size. Isaac Newton had also been aware of this problem, but suggested in the *Principia* the handy means of cometary collisions as the source of replenishment for the sun and stars. Priestley had a different argument. He used Michell's data for an elaborate approximation concerning the materiality of light and the weight loss of the sun. If, in fact, light had a definite momentum, and was considered a material particle, then its continuous emission from luminous bodies should produce a decrease in weight of the bodies which emitted light. Priestley's approximations showed that if light were material and passed approximately in the momentum Michell had detected, then the sun's semi-diameter would be shortened by about 10 feet every 6,000 years, assuming the sun's density were that of water.

The conclusions to be drawn from Michell's experiments and Priestley's calculations reinforced the scheme of Newtonian optics. The materiality of light could be admitted with full confidence, since it was calculated to be impossible to detect any weight loss as a result of this assumption. The sun's loss from emission of light was certainly too small to be detected. The notion that light was a material body with momentum fitted very well, of course, with the concept of the various sizes of colored light particles. These variously sized particles of light were affected differently by the attractive and repulsive forces of matter. The process of producing the colors could be considered as very similar indeed to the effect of variously sized visible bodies whose paths were changed by forces of attraction—precisely what Newton had suggested in the *Principia*.

If light were composed of material particles possessing momentum, however, a problem arose as to the ease with which light penetrated transparent material bodies. This was a problem which arose when light was spoken of as really material. Before Michell's experiments, it was easy enough to pass off this problem by saying that light was extremely subtle, and that bodies were extremely porous. Newton himself argued for the extreme porosity of matter. Newtonians had found it easy to believe that light, which was very "subtle," simply passed through matter, which could be very porous. But with the attempted definition of light as a material body possessing momentum, the problem of penetration of bodies became less easy to escape. Light was still very subtle, to be sure, but no longer as subtle as one desired to explain away the problem. Priestley had even estimated the weight of light in his calculations from Michell's experiment as follows:

> If we impute the motion produced in the above experiment to the impulse of the rays of light, and suppose, that the instrument weighed ten grains, and acquired a velocity of one inch in a second, we shall find that the quantity of matter contained in the rays falling upon the instrument in that time, amounted to no more than one twelve hundred millionth part of a grain, the velocity of light exceeding the velocity of one inch in a second, in the proportion of about twelve thousand millions to one, for it is nearly after the rate of two hundred thousand miles in a second. Now the light in the above experiment was collected from a surface of about three square feet, which reflecting only about half what falls upon it, the quantity of matter contained in the rays of the sun, incident upon a square foot and half of surface, in one second of time, ought to be no more than the twelve hundred millionth part of a grain, or upon one square foot only, the eighteen hundred millionth part of a grain. But the density of the rays of light at the surface of the sun is greater than at the earth in the proportion of forty five thousand to one; there ought, therefore to issue from one square foot of the sun's surface in one second of time, in order to supply the waste by light, one forty thousandth part of a grain of matter, that is a little more than two grains in a day, or about four millions seven hundred and fifty two thousand grains, which is about six hundred and seventy pounds avoirdupois, in six thousand years, a quantity which would have shortened the sun's semidiameter no more than about ten feet, if it was formed of matter of the density of water only.[31]

Priestley decided that the phenomena of the penetration of matter by light was in need of some explanation, beyond the traditional dismissal of the problem by "subtilty" and "porosity." For a solution to this problem, he turned to the work of the Jesuit priest, Roger Joseph Boscovich.

By the 1770s, Boscovich's work had become well known in England.[32] Priestley's writings suggest that Boscovich had been read and generally accepted by the Newtonians in England. The work *Theoria Philosophiae Naturalis*[33] was familiar to Englishmen and contained much that was of interest to the study of optics. Boscovich had devised a system of attractive and repulsive forces associated with matter in an attempt to explain the interaction and impact of physical bodies. His system agreed very nicely with the Newtonian system of optics, particularly in the explanations offered for reflection and refraction of light by matter. His diagrams and explanations of these phenomena bore great resemblance to those offered by Robert Smith in his *Compleat System of Opticks*.

Priestley recognized the compatibility of Boscovich's system with Newtonian optics. He saw in Boscovich's work the answer to the problem of maintaining the concept of the materiality of light. Priestley offered the following summary of Boscovich's system as it applied to Newtonian optics:

The easiest method of solving all the difficulties attending the subject of this section, and of answering M. Euler's objections to the materiality of light, is to adopt the hypothesis of M. Boscovich who supposes that matter is not impenetrable, as has been, perhaps, universally taken for granted, but that it consist of physical points only, endued with powers of attraction and repulsion, taking place at different distances; that is, surrounded with various spheres of attraction and repulsion, in the same manner as solid matter is generally supposed to be: Provided, therefore, that any body move with a sufficient momentum, to overcome any powers of repulsion that it may meet with, it will find no difficulty in making its way through any body whatever; for nothing will interfere, or penetrate one another, but *powers,* such as we know, do, in fact, exist in the same place, and counterbalance or overrule one another; a circumstance which never had the appearance of a contradiction, or even of a difficulty.

If the momentum of such a body in motion be sufficiently great, M. Boscovich demonstrates that the particles of any body through which it passes, will not even be moved out of their place by it. With a degree of velocity something less than this, they will be considerably agitated, and ignition might perhaps be the consequence, though the process of the body in motion would not be sensibly interrupted; and with a still less momentum, it might not pass at all.[34]

This argument for the ease of penetrability of light through material bodies was quite persuasive. It seemed reasonable in itself, and it also had the advantage of direct support by the eighteenth century measurements of the speed of light. The speed of light particles seemed almost incredible in their swiftness, compared to anything in the realm of eighteenth-century experience. The extreme velocity and the penetrability of light particles through matter were combined in the work of Boscovich. In his own words, he attributed the uninterrupted penetrability of light through matter as follows:

. . . Perchance that is the reason why the Divine Founder of Nature willed so enormous a velocity should be given to light, . . .

Michell's experiment and discussion of the momentum of light provided Priestley with the start of a train of argument which incorporated Boscovich's work and a support for Newtonian optics.

Priestley also turned to Michell for an attempt at a solution to the

The Construction of a System

embarrassing problem of Newton's mention of "Fits of easy reflexion and transmission" in the phenomena of the colors of thin plates and those between object glasses. Priestley supplied a rather critical and disparaging description of Newton's use of the "Fits." He also included Boscovich in his scorn because of his use, with slight modification, of Newton's concept.[35] Priestley then concluded:

> Upon the whole, therefore, it is probable, that all the mystery of these colored plates depends upon the attractions and repulsions of the particles of the bodies that compose them, affecting different rays in a different manner, according to their thickness. My objections to Newton's manner of accounting for the colors of thin plates are of long standing, but the hint of accounting for them in the manner that I have attempted to do, by the doctrine of *attractions and repulsions,* was first suggested to me by Mr. Michell, agreeably to whose conjectures relating to this subject, I have given the preceding account of the probable cause of these appearances.[36]

Priestley was convinced that the phenomena of optics should be explained by means of forces of attraction and repulsion acting upon bodies of light. In fact, he was convinced that this was also Newton's belief, apart from brief, and in Priestley's opinion, inappropriate lapses. Of Newton's optics, Priestley wrote:

> ... It never occurred to any person before Sir Isaac Newton, that reflexion and refraction may becaused by *powers of repulsion and attraction,* belonging to bodies, and extending to a certain distance beyond their surfaces. Upon this supposition, and that of light consisting of particles emitted from luminous bodies, he demonstrates, in his *Principia,* that the sine of the angle of incidence must always be to the sine of the angle of refraction in some certain ratio.[37]

Priestley sought an explanation in terms of forces of attraction and repulsion. He turned to Michell's explanation as follows:

> As to the transmission or reflexion of certain kinds of light only, producing the colours in the thin plates, the cause may be this; viz. that every particle of the medium has a great number of equal alternate intervals of attraction and repulsion, relatively to the particles of light, but that these intervals are of different magnitudes, according as the particles of light are of different colours. Now if the thickness of any transparent medium in which the particles of matter are uniformly placed, is such, that the attracting intervals of the extreme particles, as well as the repelling intervals, coincide with one another, i.e. attracting with attracting, and repelling with repelling, in regard to any one kind of rays, as the red for instance, by the united force of these extremes (all the intermediate particles of the medium mutually destroying each others effects) these rays will be reflected. But where the

plate is of an intermediate thickness, between this and the next thickness, where the attracting intervals coincide, attracting with attracting, and repelling with repelling, the attracting intervals will coincide with the repelling ones, and the repelling ones with the attracting ones, and these mutually destroying one anothers effects, these rays will pass on freely, and be transmitted. But as the intervals of attraction and repulsion are different for differently coloured rays, as has been sufficiently demonstrated, in this and the preceding section, the thickness of the plates at which these coincidences will, or will not happen, in the differently coloured rays, will be different.[38]

Priestley adopted Michell's explanation rather than Boscovich's because it was simpler and more consistent with the other assumptions of Newtonian optics. Priestley believed, with Michell, that light should have uniform properties, and that it was affected in various circumstances by the action of various forces of repulsion and attraction. This, after all, was the basic approach of corpuscular dynamics. Both Newton's and Boscovich's schemes were viewed as excessively complicated at best, and were felt to include unnecessary elements. Whatever the source of Michell's suggestions, they were in greater conformity with the major emphasis of Newtonian optics in the eighteenth century.[39]

John Michell made highly creative use of corpuscular optics in his investigations of sidereal astronomy. His work presents a striking unity of approach in the fields of optics, mechanics, and astronomy. Michell sought to find a means to measure the distance of the stars. In a paper in the *Philosophical Transactions of the Royal Society* in 1767 entitled "An enquiry into the probable parallax and magnitude of the fixed stars from the quantity of light which they afford us," he attempted to estimate the parallax of the stars by their brightness.[40] He concluded generally that the parallax of the fixed stars was less than 2" and probably less than 1" for even the closest star, Sirius. The paper was remarkable also for Michell's suggestion of the existence of binary stars sharing gravitational interaction and for his statistical treatments of large groups of stars to support his suggestion of binary stars and stars in connected gravitational systems.

Even though the topic of John Michell's 1767 paper was astronomy, his commitment to corpuscularian optics was made apparent in the last section of his paper, dealing with a possible explanation for the puzzling phenomenon of the twinkling of the stars. He offered an explanation in terms of the numbers of "particles of light" striking the eye. "It is not, I think, unreasonable to suppose," he wrote, "that a single particle of light is sufficient to make a sensible impression upon the organs of sight." He goes on to argue that the twinkling occurs because of the uneven numbers of particles of light which strike our eyes over a period of time. The star seems brighter as more particles strike our eye and dimmer as we receive less. So too, changes in color of the star can be accounted for quite nicely by changes in the number of particles in the rays of color. Variations in the density of particles in the colored rays account for shifts in the apparent color of the star.[41] This explanation is interesting primarily because of its clear reliance upon Newtonian optical assumptions.

John Michell made additional use of Newtonian optics in a paper read before the Royal Society of London in 1783. He described a new method for attempting to find the distance of the stars by a determination of the star's gravitational effect upon light particles.[42] Michell indicated to his friend Cavendish in a letter published with the paper that the whole notion of determining the "distance, magnitude, and weight" of the stars occurred to him after making some calculations on the comparison between the force of gravity and the force by which bodies emit light particles. Priestley included these calculations in his *History of Vision, Light and Colours*.[43] Michell hoped to be able to obtain data on stellar mass through the measurement of the gravitational retardation of stellar light. He had already performed the calculations for the sun's effect on light.

Michell's calculations were based upon Newton's presentation in Book I of the *Principia*.[44] He also used Newton's estimate that the force of gravity at the surface of the sun was 28 times that at the surface of the earth. After proceeding with calculations as to the velocity of light and bodies falling from infinite height to the surface of the sun, he concluded that the ratio of these two velocities was about 497 squared or 247,000 to 1. There was, therefore, a great difference between the force of gravity and the force acting upon light. The disproportion was so great that Michell expected very little diminution of the velocity of light by gravity. He went on to estimate that the ratio of the force "by which the particles of light are propelled was to gravity as about 1.9×10^{19} to 1." Michell assumed that this rather astonishing force acted over an interval of less than 1/100 of an inch. After this short-range action, the force of gravity took its effect on the motion of the particle of light. Priestley described Michell's reasoning as follows:

> From the vast disproportion between these forces it appears what an extremely little diminution the velocity of light can suffer by the attraction of the sun; the velocity at its first issuing from thence, being to the velocity diminished as much as it possibly can be by this cause, in the proportion of the square root of 247,009 to the square root of 247,008; so that the diminution of the velocity, on this account, cannot amount to more than one 492,032d part of the whole; and of this diminution, there will remain only one 45,878th part to take place, after the sun's light has arrived at the distance of the earth.[45]

Michell doubted that he would ever be able to establish measurements of so small a change in velocity. But he was encouraged into pursuing this line of thought when William Herschel's observations of double stars were published in his 1782 catalogue of multiple stars. Michell explained to Cavendish in a private letter that Herschel's discovery of a large number of new binary stars made him begin to think, "that possibly the diminution of the velocity of light might now begin to be the foundation of some observations, which might be applied to some good purpose: . . ."[46] The good purpose was the subject of the 1784 paper—to find the mass of the stars. By making assumptions about stellar distances and about the density of the stars relative to that of the sun, the measurement of the diminution in velocity could yield an indication of the stellar mass.

The procedure for the detection of the change in velocity was based upon noticing a change in the refrangibility of light. Michell took Newton's line of reasoning from the *Opticks* in which he developed the arguments for the ratio of the sine of incidence to the sine of refraction of a ray of light. Newton stated "That Bodies refract Light by acting upon its Rays in lines perpendicular to their Surfaces." Michell wrote more explicitly: "For let us suppose with Sir Isaac Newton that the refraction of light is occasioned by a certain force impelling it towards the refracting medium, an hypothesis which perfectly accounts for all the appearances."[47] Newton argued that the velocity of light could be related to the ratio of the sine of incidence and refraction as follows:

> If any Motion or moving thing whatsoever be incident with any Velocity on any broad and thin space terminated on both sides by two parallel Planes, and in its Passage through that space be urged perpendicularly towards the farther Plane by any force which at given distances from the Plane is of given Qualities; the perpendicular velocity of that Motion or Thing, at its emerging out of that space, shall be always equal to the square Root of the sum of the square of the perpendicular velocity which that Motion or Thing would have at its Emergence, if at its Incidence its perpendicular velocity was infinitely little."

Newton proceeded with mathematical demonstration to show that the sine of incidence was to the sine of refraction, "in a given ratio." He then added his usual cautionary statement:

> . . . "And this Demonstration being general, without determining what Light is, or by what kind of Force it is refracted, or assuming anything farther than that the refracting Body acts upon the Rays in lines perpendicular to its Surface; I take it to be a very convincing Argument of the full truth of this Proposition.
>
> So then, if the *ratio* of the Sines of Incidence and Refraction of any sort of Rays be found in any one case, 'tis given in all cases; and thus may be readily found by the Method in the following Proposition.[48]

The "following Proposition" was very interesting in this context because it was the proposition which John Dolland found to be in error. It was Proposition VII. Theor. VI "The Perfection of Telescopes is impeded by the different Refrangibility of the Rays of Light." Michell recognized Dolland's work. The method he devised to measure the diminution of the velocity of light was based on the use of an achromatic prism devised by Dolland.

Michell took Newton's argument about the ratio of the sines of incidence and refraction and modified it to include the effect of the action of gravity on light. If the velocity of light diminished, the ratio should change. A change in the ratio would produce a change in refraction and therefore a corresponding possibility for detection. Detection depended upon the use of one of the binary star systems listed by William Herschel and an achromatic prism. By pointing the prism at the binary star so that the rays of light are refracted first by one surface of the prism and then by the other, a difference in refraction should be detectable when comparing the light received from the larger central star and the smaller star held in gravitational (and mutual) revolution. If the stars were assumed to have the same density as the sun, a star about 22 times the sun's diameter should produce a diminution in proportion to one one-thousandth. Michell believed that an achromatic prism would be capable of revealing this small a diminution. The observational effect would be revealed as a slightly different angular separation of the two stars in the binary system in the two different orientations of the prism.

There were other problems of astronomical measurements to be solved before this prismatic data, if obtainable, could be used to determine the mass of the binary stars. The period of revolution of the stars and the angular diameter of the larger star must be determined, and an estimate of the distance made, before an actual number could be obtained. This was the first stage of the development of sidereal astronomy, with all the problems of luminosity, intrinsic brightness, distance and size yet to be worked out. In this context, Michell's suggestion for obtaining an actual measurement of a stellar parameter was met with enthusiasm.

Henry Cavendish succeeded in convincing Michell that he must overcome his reluctance to bring his work to public attention. Cavendish then related the idea to Nevil Maskelyne, the Astronomer Royal, and to William Herschel, both of whom received it with interest. Cavendish and Maskelyne were convinced, however, that an achromatic lens would be better than an achromatic prism to determine the diminution of velocity. Michell had thought of this first, but then settled on the prism as easier to work with. Cavendish calculated that a diminution of velocity as small as 1 part in 1,000 would yield a focal length change in the achromatic lens of 17 parts in 10,000. Maskelyne thought that an even smaller change in the focal length could be detected.

Nevil Maskelyne also tried to detect a change in the refrangibility of light by using an achromatic lens to observe some stars. Herschel, too, tried the same observation on a "great many stars." Both men could not detect any change, so Herschel set to work grinding a prism to try the observations as Michell had suggested. Cavendish concluded, in a letter to Michell, that as a result of the failure of Maskelyne and Herschel "there is not much likelyhood of finding any stars whose light is sensibly diminished." Cavendish seemed satisfied that the theory was correct, but that the magnitude of the effect was too small to detect. The sizes of the stars were not known and he thought that there were probably no stars large enough to produce a detectable diminution.[49]

Michell was somewhat disappointed by these negative findings. He acknowledged Cavendish's conclusion that it was "very possible there may be no stars large enough to produce any sensible effect," but he also conceived of another alternative; that

> it is also possible, that light (and perhaps too the electrical fluid, which seems to be in some degree allied to it, etc.) may not be so much affected by gravity, in proportion to their vis inertia, as other bodies; but though I am much more inclined to believe this is not the case, yet the singular properties those substances are possessed of, seem to leave little more room for doubt with respect to them, than other common kinds of matter.[50]

Michell was willing to entertain the thought that perhaps gravity affected light and the electrical ether differently—probably because of the discouragement he felt over the failure of his plan. His suggestion had been based on the extension of the uniformly acting force of gravity to the whole of the universe: to the stars, to binary stars, and to the action of gravity on light as it traversed interstellar space. Michell would not have been the astute natural philosopher that he was, had he not considered the possibility of questioning his basic assumption, in view of the negative observational results. This was especially true in the context of the eighteenth-century enthusiasm for imponderable fluids. But both Michell and Cavendish were not inclined to give much credence to these imponderable fluids because of their commitment to a Newtonian worldview of their own devising—a world-view consisting of forces and ponderable matter.

Both John Michell and Henry Cavendish took very seriously Newton's statements in both the *Opticks* and *Principia* concerning the importance of the investigation of forces and corpuscular explanations. Newton had provided sufficient information for them to feel secure that the approach they had developed for the investigation of the physical world was in conformity with Newton's program as outlined in his published works. In Newton's Preface to the first edition of the *Principia,* he outlined the following program for investigating the System of the World:

... In the third Book I give an example of this in the explication of the System of the World; for by the propositions mathematically demonstrated in the former Books, in the third I derive from the celestial phenomena the forces of gravity with which bodies tend to the sun and the several planets. Then from these forces, by other propositions which are also mathematical, I deduce the motions of the planets, the comets, the moon, and the sea. I wish we could derive the rest of the phenomena of Nature by the same kind of reasoning from mechanical principles, for I am induced by many reasons to suspect that they may all depend upon certain forces by which the particles of bodies, by some causes hitherto unknown are either mutually impelled towards one another, and cohere in regular figures, or are repelled and recede from one another. These forces being unknown, philosophers have hitherto attempted the search of Nature in vain; but I hope the principles here laid down will afford some light either to this or some truer method of philosophy.[51]

The clear implication, which could be derived from both the *Principia* and the *Opticks* if one were selective, was that the whole of Nature was explicable in terms of mathematical laws, descriptive of the action of forces. Because of Newton's success in the treatment of the solar system, the moon, comets, and the seas, his followers—Robert Smith, Roger Cotes, most of the popularizers of his works, and John Michell and Henry Cavendish—believed that the rest of Nature would be amenable to the same approach. It was successful in Newtonian optics and Michell and Cavendish hoped it would be equally successful in sidereal astronomy.

Cavendish, especially, seemed committed in his scientific works to Newton's suggested program—as Russell McCormach has argued persuasively:

It was the conception of natural philosophy as the search for the forces of particles, exemplified in the texts of the *Principia* and *Opticks* and discussed in the *Principia*'s preface and the *Opticks*'s queries, which guided Cavendish's scientific explorations. His various and seemingly disconnected researches were truly aspects of the same investigation, for there was only one basic problem: to discover the forces of particles. The manifold nature of his work was not a mystifying idiosyncrasy but a reflection of the richness of detailed illustration—from the deliquescence of salt of tartar and the balancing of a fly on water to the eruption of a volcano—of the action of attracting and repelling forces of Newton's thirty-first query. Cavendish was a close student of the *Principia*—forever his model of quantitative science—for, it seems, the queries' plea for the universal explanatory power of Newton's corpuscular philosophy was entirely convincing to him.[52]

This was the orientation toward the study of nature which made the unity of approach of Newtonian optics, mechanics, and astronomy so natural.

Henry Cavendish also attempted his own variation of applying the concept of gravitational attraction to light. He calculated the "bending of a ray of light which passed near the surface of any body by the attraction of that body." He thought perhaps the observation of the bending of light might be an alternative method of deriving data by which to weigh the stars, since the diminution of velocity proved undetectable.[53]

Both Michell's and Cavendish's attempts to apply Newtonian optics to sidereal astronomy represent an interesting example of the operation of a world-view. Their conviction about the nature of the universe was such that failure to produce observational results did not dissuade them from believing that their fundamental scheme of the interaction of light and matter was correct. It made excellent sense in the context of their extension of force concepts, derived from Newton's published works, to the whole of the universe. They were attempting to make explicit what Newton had only hoped for in the *Principia* and applied to the solar system: the universality of gravitational attraction.

William Herschel, too, regarded Newton's program for the investigation of forces very seriously. He was steadfast in his belief in the actions of central forces upon bodies, be they gross matter or light. Herschel derived information concerning various schemes of the relationship of forces to matter from his reading of Priestley's *History of Vision, Light and Colours*. He knew of what he called "Father Boscovich's and Mr. Michell's hypothesis" at second hand from Priestley, and he devised his own scheme of central forces along lines rather distinct from what he understood their hypotheses to be. Herschel took some care in attempting to convince the scientific society to which he belonged—the Philosophical Society of Bath—that his own conception of central forces differed from that of both Boscovich and Michell. He thought that these implied the immateriality of matter, but, Herschel declared, "I ascribe the *central powers* to *particles of matter*: Divisible *ad infinitum*; *impenetrable*; in short—material, in the usual acceptation of the word."[54]

Herschel developed his scheme of central forces at some length, arguing for the existence of central forces from the evidence of observations and experiments. He did not agree with the Boscovichian scheme of zones of forces about points to describe matter because he

The Construction of a System

believed these zones of attraction and repulsion to be unchanging in their strength until—for example—suddenly they changed completely from attraction to repulsion. In his own scheme, the forces acted according to the reciprocal of distance, raised to some high, but not exactly specified power. He described his scheme to the Philosophical Society of Bath as follows:

... My Hypothesis, then, is that every particle of matter, is endowed with at least four more *central powers*, each of them acting under very different Laws; but none of them under the law of that power, established by the newtonian philosophy.

It will presently appear that this Hypothesis can solve every difficulty of our question; And if, by a careful investigation of nature, and attention to experiments, we could arrive at the knowledge of the laws by which these powers act, we should be able to show, *a priori*, in what manner a ray of light is stopped. But, to proceed, I shall beg leave to call these powers by the following names.

1)—The attraction of cohesion.
This will be found to be in the inverse ratio of some very high power of the distance.

2)—The repulsion of emission or reflexion of light. (I do not mean that this power is only to serve for the purpose of emitting or reflecting light but only that it is adequate to it.) This power must also be in the inverse ratio of some high involution of the distance, but far inferior to the former.

3)—The attraction of refraction and inflection.
This again, must be in the inverse ratio of some considerable power of the distance.

4)—The repulsion of deflection.
Still in the inverse ratio of some power higher than the square of the distance.

5)—The Newtonian attraction of gravity.
Known to be in the inverse ratio of the square of the distance.

As I observed before, every one of these central powers must act according to different laws. For should any one repulsive power act under the same law with an attractive force, they would exactly balance each other's effects, and thereby destroy one another. And should two powers of the same kind act under the same law, it would be no more than doubling the force; that is to say no more than one power.

It is evident at first sight that this Hypothesis is fruitful beyond all bounds, since from the different number and connections of particles tho' possessed of no other central powers (which however I would by no means affirm) such very different effects, in regard to the extent of the circles or spheres of action must arise as will evidently carry us into an endless field of speculation and experiments. And for this very reason the investigation of the laws under which these central powers act will be attended with the utmost difficulty, and require all the assistance that mathematics joined to experimental philosophy can afford.[55]

Herschel's description of forces was made in response to what he called Dr. Priestley's Optical Desideratum—"What becomes of Light?" in January 1780. He concluded his discussion with what he considered to be an excellent support for his own scheme of forces, including an interesting drawing (see Fig. 4):

> Dr. Priestley's Desideratum mentions also, that, "if the same cause produce all these effects, it might be expected that it would sometimes produce effects of an intermediate kind; in consequence of which light might not always be reflected or refracted with equal velocity."

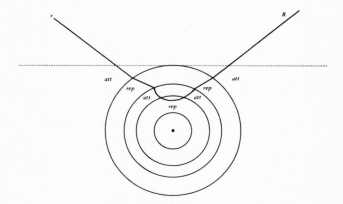

Fig. 4   Light acted upon by forces.

> In answer to this, it follows most evidently from my hypothesis of central attracting and repelling powers that no such intermediate effect can ever take place. For if a particle of light can possibly escape at all either by reflexion or refraction, it must come off with the same velocity it enter'd the medium, or came within the sphere of its active powers; because there will always be a perfect ballance of attraction to attraction, and repulsion to repulsion in the same sphere. The bare inspection of the annexed figure will be a sufficient proof of what I advance, which however I could easily give in a mathematical form if it were required.[56]

William Herschel's work in optics is interesting primarily as an example of acceptance of the assumptions of Newtonian optics.[57] His contribution to physical optics certainly did not compare to his work in geometrical optics, lens grinding, and telescope building, but his orientation seemed consistent. He derived his information on physical optics from secondary works, primarily from Priestley's *History,* and from Smith's *Compleat System of Optics,* which he held in high regard. His belief in the explanation of optical phenomena in terms of forces acting upon light was clearly compatible with his work in astronomy. His enthusiasm over Michell's suggestion for the measurement of the diminution of the velocity of light demonstrated his acceptance of the optical assumptions behind Michell's scheme.

Support for Newtonian optics came from several other quarters as the eighteenth century drew to a close. In 1788 the Royal Society of Edinburgh heard a paper by John Robison, Professor of Natural Philosophy at the University of Edinburgh, which argued strongly for the Newtonian scheme.[58] Robison undertook the paper because he detected an error in Boscovich's suggestion that an observer could detect the motion of the earth by making observations with a telescope filled with water. His paper, therefore, undertook to analyze Boscovich's reasoning, and point out its error. In the process, however, he became interested in a "hitherto unconsidered subject in physico-mathematical science, *the motion of light as affected by bodies which are also in motion."* The last part of his paper was devoted to the determination of the reflection and refraction of light from moving surfaces. Robison believed that light consisted of corpuscles emitted from shining bodies, and not undulations of an elastic fluid. He was convinced that refraction was produced by forces acting perpendicularly to the refracting surface.

Robison took an interesting philosophical position with respect to Newtonian optics in this paper. He chose to stress that Newtonian optics could be demonstrated from the phenomena. Robison interjected a strong empirical element into his support for Newtonian optics. His commitment to Newtonian assumptions was clear from the first part of his paper. But he feigned objectivity in his choice of optical principles when he decided

to take up such opinions concerning the nature of light, as seem most rationally deducible from the phaenomena which we observe, and then to deduce, by the established principles of mechanics, such consequences as should arise from the action of refracting and reflecting substances upon this hypothetical light. We should then select such of these consequences as will admit of a comparison with observation. If these consequences shall be found inconsistent with observation, the hypothesis concerning the nature of light must be rejected, and trial must be made of a new one. But if they should be found to agree with observation, and at the same time be sufficiently various, we may then admit the hypothesis to have a degree of probability proportioned to the extent of the comparison which we have made of its consequences with observation; we may then discover by this means parts of a hypothesis which must be admitted as true, although the hypothesis cannot be demonstrated in its full extent.[59]

Robison claimed that there were only two hypotheses concerning the mechanical nature of light, which seemed to be rationally deduced from the phenomena. There was that which was "advanced by Sir Isaac Newton, in several parts of his celebrated writings," and "the other hypothesis" of "Mr. Huygens and Dr. Hooke." Robison described Newton's optics in strict Newtonian terms, involving light particles acted upon by attractive and repulsive forces. When conclusions were drawn from this hypothesis, and the results compared with observation, "the most perfect agreement is still discovered. For these reasons, this hypothesis had acquired great credit, and deserves to be examined on the present occasion." The rival hypothesis did not fare as well:

The other hypothesis is that of Mr. Huyghens and Dr. Hooke. These gentlemen suppose that, as hearing is produced by means of the tremulous motion of elastic air, which affects the ear, so vision is produced by the tremulous motion of elastic light, which affects the eye. This hypothesis was announced and applied to the explanation of phaenomena in very general terms, and did not, for a long while, much engage the attention of the learned. The celebrated mathematician Mr. Euler has lately brought it into credit, having made some alterations in it. He supposes, that vision is produced by the tremulous motion of an elastic fluid which he calls aether, and which he supposes to pervade all bodies. He attempts to show that the propagation of this tremulous motion is analogous to the appearances in the reflection and refraction of light. I confess that I cannot admit his reasonings on this subject to be agreeable to the principles of mechanics; and I am decidedly of opinion, that the propagation of the tremulous motion of an elastic fluid is totally inconsistent with those facts in vision where no refraction or reflection is observed. But I shall reserve my objections till another opportunity, when I propose to submit to this Society a mechanical examination of this hypothesis, and I shall admit for the present that Mr. Euler's explanation of refraction and reflection is just. It is an essential proposition in this hypothetical theory, that the velocities of the incident and refracted light are proportional to the sines of incidence and refraction, and therefore that light is retarded when it is refracted toward the perpendicular. It seems a necessary consequence that, in this case, the particles of aether are actuated by forces tending from the refracting body. I shall, therefore, consider what effects must result from the combination of this retardation with the motion of the refracting body. If time will allow, I shall consider what will be the effects produced on the motion of light by the motion of the visible object. These are so different in the two hypotheses, that it is very probable that some natural appearance may be found which will give us an opportunity of determining whether either of these hypotheses is to be received as true.[60]

Robison was entirely correct in his suggestion that the two hypotheses could be subjected to experimental comparison if it could be determined

The Construction of a System

whether light were accelerated or retarded upon entering a medium of higher index of refraction. But this experimental determination did not occur until after 1849, with the work of Fizeau and Foucault. Until then, the choice between the two hypotheses remained largely a matter of personal preference. It was clear that Robison opted for the Newtonian system.

The calculations which Robison performed were styled after Newton's geometrical demonstration in the *Principia*. They were all founded on a basic proposition, from which all the conclusions were drawn, namely:

> If a ray of light, moving in any direction and with any velocity, meet with the surface of a refracting medium, which is in motion, its final relative motion will be the same as if the medium had been at rest, and the light had approached it with the same initial relative motion.[61]

Robison's was a detailed and convincing attempt to extend the Newtonian explanation for the reflection and refraction of light to the more generalized case of media in motion. Previous to this, concern had been limited to the more specialized situation of stationary reflecting and refracting substances. It is interesting to note that Robison described his work to John Playfair at the beginning of 1784, and Playfair devised an "elegant analytical demonstration," supporting Robison's geometrical demonstration with the use of the method of fluxions. Robison included in his own paper Playfair's calculation of "the velocity of a particle of light" as it entered the zone of action of the force outside a moving refracting medium. The result of both geometrical and analytical demonstrations was the extension of Newtonian optics to moving media which reflected and refracted light. The relative velocities of the incident and refracted and reflected light particles were calculated and their changes in path were geometrically represented according to the motion of the media. There was, of course, no reason to suspect that this application of the so-called Galilean transformation in mechanics to optics was incorrect.

Robison was able to put his calculations to practical use, because he was able to demonstrate that Boscovich's suggestion that a water telescope would exhibit diurnal deviation was incorrect. His extension of Newtonian optics allowed him to solve a group of observational problems of current interest concerning the use of the telescope and the measurement of the aberration of light. He concluded that a "water telescope must have the same position with the common telescope, or that both of them must always be directed to the real place of the terrestrial object." He also concluded that a water telescope and a common telescope would both indicate that same aberration of the fixed stars. The assumptions of Newtonian optics proved themselves capable of extension to meet problems in optics of current concern.[62]

By the last decade of the eighteenth century, Newtonian optics rested upon a basis of wide acceptance and convincing explanation. But there were still the difficult areas of inflexion (diffraction) phenomena and the colors of thin plates to be brought completely into the system. The young Henry Brougham believed that he had discovered the means of bringing these recalcitrant phenomena into the fold in the 1790s. Brougham rushed in where all other Newtonians feared to tread in his attempt to explain these phenomena. His brief career in optics is interesting both in terms of its conformity to Newtonian optical assumptions and in terms of its superb illustration of how cumbersome Newtonian optics had become by the end of the century.

Brougham was clearly committed to an explanation of optical phenomena in terms of forces and particles. He was convinced that there were zones of forces outside every medium which acted upon the differently sized particles of light represented by the various colors. Henry Brougham was the most elaborate example of a Newtonian optician in the eighteenth century. Ironically, he believed that his work was in close agreement with the optical works of Isaac Newton, in their purest form. It is clear, however, that what Brougham accepted as Newton's optics was very different from Newton's works and especially from assumptions that Newton would have found acceptable. Brougham's two optical papers of 1796 and 1797[63] were in many ways similar to what had become the standard Newtonian interpretation of Newton's optics. Several passages from Brougham's papers will clearly show his Newtonian position:

That bodies reflect light by a repulsive power, extending to some distance from their surfaces, has never been denied since the time of Sir Isaac Newton. Now this power extends to a distance much greater than that of apparent contact, at which an attraction again begins, still at a distance, though less than that at which before there was a repulsion, as will appear by the following demonstration which occurs to me, and which in general with respect to the theory of Boscovich.[64]

These observations enable us to give a very short summary of optical science. When the particles of light pass at a certain distance from any body, a repulsive power drives them off; at a distance a little less, this power becomes attractive; at a still less distance, it again becomes repulsive; and at the least distance, it becomes attractive as before, always acting in the same direction. These things hold whatever by the direction of the particles; but if, when produced, it passes through

The Construction of a System 87

the body, then the nearest repulsive force drives the particles back, and the nearest attractive force either transmits them, or turns them out of their course during transmission. Further, the particles differ in their dispositions to be acted upon by this power, in all these varieties of exertion when varied, except in the cases of refraction, of which we before spoke; and these dispositions of the parts are in all cases in the same harmonic ratio. Lastly, the cause of these different dispositions is the magnitude of the particles being various.[65]

Brougham was determined to explain all of optics in terms of forces of attraction and repulsion, within a framework of what he believed to be the true, empirical model of Newton's *Opticks*. This model allowed absolutely no place for an ether. Brougham took the traditional reluctance on the part of Newtonian opticians to discuss the ether one step further; he was aggressively opposed to any reference to ether in optics. Newton's legacy of the *Opticks* definitely did not include the ether, in Brougham's view. Writing of the phenomenon of thin films, he stated his distaste for things which he considered to be unworthy of the Newtonian heritage.

The celebrated discovery of Newton, that these [colors] depend on the thickness of their parts, is degraded by a comparison with his hypothesis of rays and waves of aether. Delighted and astonished by the former, we gladly turn from the latter; and unwilling to be involved in the smoke of unintelligible theory so fair a fabric founded on strict induction, we wish to find some continuation of experiments which may relieve us from the necessity of the supposition.[66]

Brougham not only wanted to maintain the Newtonian heritage in what he regarded as its purest form, he wanted to extend optics by offering a new explanation for the phenomena of the colors of thin and thick films and striated surfaces. These phenomena had presented the most difficult problems to the Newtonians who sought to avoid speaking of "Fits of easy transmission and reflexion" and the ether. Brougham was convinced that he had devised a new explanation of the colors of films and striations, using force zones, particles, and a new property which he assigned to light—the various reflexibilities of the different colors of light. He was sure that light corpuscles had different reflexibilities according to their size and color.

It has always appeared wonderful to me [he wrote], since nature seems to delight in those close analogies which enable her to preserve simplicity and even uniformity in variety, that there should be no dispositions in the parts of light, with respect to inflection and reflection, analogous or similar to their different refrangibility. In order to ascertain the existence of such properties, I began a course of experiments and observations. . . .[67]

Brougham was convinced that the problems of colors of films and striated surfaces could be explained in terms of differently sized light corpuscles, possessing different reflexibilities, and acted upon by zones of force. He proposed that the various particles of light did not all have the "same disposition to be acted upon by bodies which inflect and deflect them."[68] Brougham further assigned a physical cause to the reflexibility:

As light is reflected by a power extended to some distance from the reflecting surface [he wrote], the different reflexibility of its parts arises from a constitutional disposition of these to be acted upon differently by the power. And as these parts are different sizes, those which are largest will be acted upon most strongly.[69]

The power extending outside the reflecting surface was believed to vary in a uniform way, in a manner similar to $1/r^2$, though not necessarily gravity. This was a natural assumption, using Newton's treatment of forces as a model. It was also a necessary assumption, for Brougham wanted to use the theorems from Book I of the *Principia* to actually derive the equation of the paths of the bodies of light in the force zones. If the paths varied, as Brougham was sure they did, then the cause must be in the light particles themselves. If the force on the particle could be found, a comparison of the paths of the different light particles could yield their relative sizes.

Using the properties of reflection, Brougham represented the actual force acting upon a given light corpuscle by the equation

$$F = \frac{V \times \sin(R + I)}{\sin R}$$

where V = velocity, R = the angle of reflection, and I = the angle of incidence.[70] This enabled Brougham to determine the exact sizes of the light particles responsible for the color phenomena.

... The force exerted on the red [he wrote] is to that exerted on the violet, as the size of the red to the size of the violet (by hypothesis), therefore, the red particles are to the violet as 1275 to 1253. . . .

... All this follows mathematically, on the supposition that the parts of light are acted upon in proportion to their sizes; and to say the truth, I see no other proportion in which we can reasonably suppose them to be influenced; for such an action is conformable to the universal laws of attraction and repulsion. . . .[71]

Brougham's work in optics was based on the assertion that the particles of light were of different sizes. These sizes were now not only assumed to be different as they had been before, they were *calculated* to be

The Construction of a System

different. Having established the relative sizes of light particles to his satisfaction, Brougham attempted to define the positions of the zones of force which acted upon light to produce the various optical phenomena. Since he assumed that the action of the force was uniform, he concluded that the various zones of force must be ordered outside a body according to their distance from the surface. These zones of force would then operate on the bodies of light according to their size. Brougham was confident that he had provided an exact, quantitative manner of accounting for optical phenomena in terms of sizes of bodies of light and the distance of the zone of force above the surface of a body. After describing a great many experiments which lent themselves to his explanation, Brougham wrote:

I infer then from the whole, that different sorts of rays come within the spheres of flexion, reflexion and refraction, at different distances, and that the actions of bodies extend farthest when exerted on the most flexible. It may perhaps be consistent with accuracy and convenience to give a name to this property of light; we may therefore say that rays of light differ in degree of refragity, reflexity, and flexity, comprehending inflexity and deflexity. From these terms (uncouth as, like all new words, they at first appear) no confusion can arise, if we always remember that they allude to the degree of distance to which the rays are subject to the action of bodies. I shall only add an illustration of this property, which may tend to convey a clearer idea of its nature. Suppose a magnet to be placed so that it may attract from their course a stream of iron particles, and let this stream pass at such a distance that part of it may not be attracted at all; those particles which are attracted may be conceived to strike on a white body placed beyond the magnet, and to make a mark there of a size proportional to their number. Let now another equal stream considerably adulterated by carbonaceous matter, oxygen, etc. pass by at the same distance, and in the same direction. Part of this will also be attracted but not so far from its course, nor will an equal number be affected at all; so that the mark made on the white body will be nearer the direction of the stream, and of less size than that made by the pure iron. It matters not whether all this would actually happen, even allowing we could place the subjects in the situation described: the thing may easily be conceived, and affords a good enough illustration of what happens in the case of light.

Pursuant to the plan I before followed, I now tried to measure the different degrees of reflexity, etc. of the different rays; but though the measurements which I took agreed in this, that the red images were much larger than the rest, and the green appeared by them of a middle size, yet they did not agree well enough (from the roughness of the images, and several other causes of error) to authorize us to conclude with any certainty "that the action of bodies on the rays is in proportion to the relative sizes of these rays." This however will most probably be afterwards found to be the case: in the mean time there is little doubt that the sizes are the cause of the fact. . . .[72]

This was a clear example of the extent to which Brougham was committed to the correctness of his assumptions about the bigness and smallness of the corpuscles of light.

The real purpose of Brougham's concern for the process of reflection and the various reflexibilities of the bodies of light finally became clear when he considered the problem of the colors of thin films and plates. Brougham saw in his ideas about reflection the answer to this troublesome problem. The colors were explained in terms of zones of force and reflection, rather than by allusion to "Fits of Easy Transmission and Reflection," which involved the unacceptable ether. Brougham described several experiments he performed in which he used the properties of reflection from various surfaces to obtain color effects. The patterns of colors he obtained by random experimentation with reflecting surfaces happened to be similar to the pattern of the colors produced in the phenomena of thin films. Confident that he had at last found the explanation, Brougham wrote:

To say that they [the color rings] are formed by the thickness of the plates is not explaining the thing at all. It is demanded, in what way? And indeed we see the like dark intervals and the same fringes formed at a distance from the bodies by flexion, where there is no plate through which the rays pass. The state of the case seems to be this: "When a phenomenon is produced in a particular combination of circumstances, and the same phenomenon is also produced in another combination, where some of the circumstances before present are wanting; we are entitled to conclude, that the latter is the more general case, and must try to resolve the other into it."[73]

Brougham argued that the images produced by the reflection experiments he had performed were similar in pattern and arrangement of colors to the colors of thin films. Therefore the phenomena of thin films must be explainable in terms of the more general phenomenon of color production by reflection. The production of colors by reflection was, of course, considered to be perfectly explained in terms of forces and light corpuscles.

Brougham was satisfied that the problem of thin films was explained in accordance with the Newtonian system of optics. He extended this imagined triumph to include the new problem of the colors of striated surfaces. The colors produced by striated surfaces and scratches were considered particularly important because he argued that many of the color patterns which men had previously linked with the problem of thin films arose simply from the striations and scratches produced on glass by polishing. His explanation of the colors produced by these striated surfaces

The Construction of a System 91

(see Fig. 5) was an excellent example of his proposed process of color formation by reflection:

> The manner in which the first of these propositions is demonstrated a priori is evident from the 4th figure, where *DC* is the reflecting surface, to a concavity bearing a small ratio to *DC*, *Ao* and *AB* rays proceeding to *DC*. The one, *AB*, will be separated into *Br* red, *Bv* violet, by deflexion from *o*, and will be reflected to *r'v'* forming there the fringes. The other, *Ao*, being reflected, will be separated into *B*λ and *By*, by deflexion from *v* forming other fringes, *xy* on the side of *vo*'s shadow opposite to 2*r'v'*. Also when *vo* is convex instead of concave the like fringes will be produced by the rays being deflected in passing by its sides. Lastly, when *vo* is a polished streak images by reflexion will be produced, as described. . . .[74]

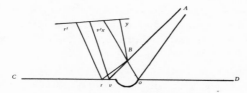

Fig. 5 Colors from striated surface.

This explanation provides us with a good example of the complexity of the system. The deformation in the surface produced a corresponding deformation in the contour of the force zones. These deformed zones acted upon the ray of light in such a way as to produce the colors. Brougham wisely did not attempt to trace a single ray in detail through the whole process. He obviously implied it could be done, however, using the appropriate curve as the ray entered each of the zones.

Brougham felt that he had performed a great service to the Newtonian tradition of optics. He had, in fact, extended the theory to cover the new phenomena of colors of striated surfaces. He had also re-explained the phenomena of thick or thin films and plates using color production by reflection. By doing this, Brougham extended the system of forces and particles to include a more detailed explanation of the phenomena. In doing this, he believed he had eliminated the necessity of ever again mentioning the hypotheses of "Fits of easy transmission and reflexion" and the ether.

By the close of the eighteenth century, the Newtonian system of optics had become a completed structure, offering explanation as well as accurate description of all known optical phenomena. The system also contained clearly visible preferences and prejudices. The works of Sir Isaac Newton were revered, as was the master himself. Forces and light corpuscles were considered appropriate concepts to use. Reference to an ether and to "Fits of easy transmission and reflexion" were considered to be decidedly inappropriate. These latter concepts were explicitly purged from the system by Brougham, although they had been viewed with disfavor by Smith and had been all but eliminated in the 1750s. Newtonian optics had coalesced into a system during the eighteenth century, but, by the close of the century, its disadvantages had also become apparent. The system was by far the most complete way to treat optical phenomena, but it was cumbersome, with its forces of attraction and repulsion and variously sized particles. The major successes of Newtonian optics in the last half of the century—if they could be called major—were "after-the-fact" explanations of previously recognized phenomena. It had become a closed system which failed to indicate new directions of investigation and interpretation.

The extent of rigidity of the Newtonian system will become quite clear as we investigate the conflict between Newtonian optics and the new wave theory of light which developed during the nineteenth century. It is important first, however, to point out that the undulatory theory of light also had a continuing though less successful tradition during the eighteenth century. This rival system must be considered if the challenge which the undulatory theory presented to the Newtonian system in the opening years of the nineteenth century is to be fully understood.

The Rival Undulatory Theory in England

An undulatory theory of light was available to English natural philosophers as early as the 1660s. The main sources for this undulatory theory were the works of the Englishman Robert Hooke and of the Dutchman Christian Huygens. The undulatory theory was set against a background of Cartesian physics and so represented both an alternate way of explaining optical phenomena and a different conception of the physical world.[75]

Robert Hooke was the most important early proponent of the undulatory theory in England.[76] He was well known to his contemporaries as an excellent experimenter and competent natural philosopher. He is unfortunately best known now as the first person to raise serious objections to the optical papers of the young Isaac Newton. Newton's new theory of light and colors was viewed as an hypothesis which challenged his own well-formulated theory of optics. Hooke devised an undulatory theory which adopted many aspects of the Cartesian concept of the ether. But he

The Construction of a System

modified Descartes's optical works, especially with regard to the theory of color formation and the instantaneous transmission of light. Hooke conceived of light in terms of a pulse moving through the ether, with a wave front whose obliquity determined the color of the light. The pulse traveled with variable speeds, according to the medium traversed. Modifications of the light produced the colors.

Hooke described light as consisting of a pulse which was produced by the motion of bodies. He wrote in his *Micrographia*:

. . . For Light, it seems very manifest, that there is no luminous Body but has the parts of it in motion more or less. . . .[77]

It would be too long, I say, here to insert the discursive progress by which I inquired after the properties of the motion of Light, and therefore, I shall only add the result.
   And, first, I found it ought to be exceeding quick, such as those motions of fermentation and putrefaction, whereby, certainly, the parts are exceeding nimbly and violently moved. . . .
   . . . Next, it must be a vibrative motion. . . .
   . . . And thirdly, that it is a very short vibrating motion. . . .[78]

On the question of propagation of light, Hooke mentioned the essential aspects of an undulatory system.

That the motion is propagated every way through an Homogeneous medium by direct or straight lines extending every way like Rays from the center of a Sphere. (Fifthly), in an Homogeneous medium, this motion is propagated every way with equal velocity, whence necessarily every pulse or vibration of the luminous body will generate a Sphere which will continually increase, and grow bigger, just after the same manner (though indefinitely swifter) as the waves or rings on the surface of the water do swell into bigger and bigger circles about a point of it, where, by the sinking of a Stone the motion was begun, whence it necessarily follows, that all parts of these spheres undulated through an Homogeneous medium cut the Rays at right angles.[79]

Hooke accepted the concept of Descartes's ether as his homogeneous medium. It was a particulate fluid whose main property was the transmission of pulses. He gave few details about the ether, except that it was "so exceeding fluid a body, it easily gives passage to all other bodies to move to and fro in it."[80]

Hooke did not, however, accept Descartes's description of light. It did not propagate instantaneously through the plenum of ether particles, nor were the colors produced the way Descartes claimed. "The case therefore of the generation of color must not be what Descartes assigned, namely, a certain rotation of the globuli etherei, which are particles which he supposed to constitute the 'Pellucid medium'. . . ."[81] Rather, Hooke set down a system whereby the obliquity of the wave front of the pulse would determine the color. He had the Aristotelian notion that red was a stronger color than blue.

> That Blue is an impression on the Retina of an oblique and confus'd pulse of light, whose weakest part precedes, and whose strongest follows. And, that Red is an impression on the Retina of an oblique and confus'd pulse of light, whose strongest part precedes, and whose weakest follows.[82]

His explanation of the refraction of light was based on the difference in the refracting media, considering that "all transparent mediums are not Homogeneous to one another." By considering the front of the pulse of light as it entered the new medium, Hooke concluded that the front of the pulse changed obliquity as it entered the new medium. His description of the process was, unfortunately, unclear and he offered no geometrical description to make his statements more exact.

Several of the important rudiments of the undulatory theory of light were found in Hooke's *Micrographia*. He was the first to describe light as a pulse with a front, advancing through a homogeneous medium with a uniform speed and changing direction as a result of encounter with different media. His work represented a partial break with the Cartesian theory of light by denying that theory's concept of color production. But he still considered the phenomena of colors to arise from a modification of white light. His work was an important start, but it was incomplete and often unclear. It further lacked a firm basis in connected experiments. Newton's early experimental work cast immediate and very serious doubt upon Hooke's whole scheme of light pulses in an ether.

The other major source of an undulatory theory of light in England came from the works of Christian Huygens. He adopted views very similar to Hooke's in his *Treatise on Light*, written in 1678.[83] His ideas on the nature of light and the ether were very close to Hooke's observations IX and XV, which we considered earlier. Light consisted of the rapid motion of some sort of matter. This motion was a mechanical motion of the particles of matter. A quotation from Huygens will show the similarity:

> It is inconceivable to doubt that light consists in the motion of some sort of matter. For whether one considers its production, one sees that here upon the earth it is chiefly engendered by fire and flame which contain without doubt bodies that are in rapid motion, since they dissolve and melt many other bodies, even the most solid; or whether one considers its effects, one sees that when light is collected, as

by concave mirrors, it has the property of burning as a fire does, that is to say it disunites the particles of bodies. This is assuredly the mark of motion, at least in the true philosophy, in which one conceived the causes of all nature effects in terms of mechanical motions. This, in my opinion, we must necessarily do, or else renounce all hopes of even comprehending anything in physics.[84]

Light, for Huygens, was explained in terms of mechanical motion. This motion was considered as taking place in the ether. To illustrate this motion, Huygens made use of the familiar concept of sound in air, but he did not, however, use an analogy between light in an ether and sound in air. He illustrated the concept of light in ether by pointing out the differences between light phenomena and the observable phenomena of sound in air, quite a different approach in this respect from those of Malebranche and Euler, as we shall see. Rather than relying upon an analogy, Huygens emphasized important differences between the two phenomena. Stating that light spread in spherical waves, as did sound, he continued:

But if the one resembles the other in this respect they differ in many other things; to wit, in the first production of the movement which causes them; in the manner in which the movement spreads; and in the manner in which it is propagated.[85]

Sound originated from the movement of the whole body producing the sound. But light must be produced by the movement of each point of the surface of the luminous body, or the whole body would not be visible. Further, the vibration of the bodies which produce sound did not produce light, "since we do not see that the tremors of a body which is giving out a sound are capable of giving rise to light, even as the movement of the hand in air is not capable of producing sound."[86]

Huygens's theory of light required an ether to transmit the undulations. He was a firm believer in the type of particulate ether proposed by Descartes. Rather than accept the various criticisms of the ether, Huygens wrote in favor of Descartes's ether: "It being on the contrary quite credible that it is this infinite series of different sizes of corpuscles, having different degrees of velocity, of which Nature makes use to produce so many marvellous effects."

Huygens's considerations on the ether were detailed. His treatment of the propagation of light and the motion of the ether was similar to his earlier work on impact physics and mechanics. He started by assuming perfect "springiness" for each particle of the ether. Therefore, an impulse communicated to a particle of ether by impact would be perfectly transmitted, maintaining the uniform velocity of light. Huygens did not envision the problem in terms of the propagation of a narrow beam of light, but rather in terms of the propagation of the individual spherical waves originating at every point on the surface of a luminous body. Huygens's work on motion through fluids and his concept of the transmission of light were based on the concepts of pulses and waves in the ether. The mechanism for the transmission of light stressed the concept of a spherical wave traveling through the fluid. He was concerned with the behavior of the wave front as it underwent changes, to produce the various optical phenomena. In fact, his system for the propagation of light was well-suited to these spherical waves because it did not concern itself with rays of light or narrow beams, as did the Newtonian system. It appears probable that Huygens did not consider the case of narrow beams at all, being instead too occupied with the concept of spherical waves as they applied to optical phenomena.

It will be useful to quote Huygens's words on both the propagation and non-confusion of light in the ether, not only for interest here, but as a basis for understanding the work of Thomas Young in the early nineteenth century. (See Fig. 6.)

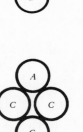

Fig. 6   Communication of an impulse.

And it must be known that, although the particles of the ether are not ranged thus in straight lines, as in our rows of spheres, but confusedly, so that one of them touches several others, this does not hinder them from transmitting their movement and from spreading it always forward. As to this, it is to be remarked that there is a law of motion serving for this propagation, and verifiable by experiment. It is that when a sphere, such as *A* here, touches several other similar spheres *CCC*, if it is struck by another sphere *B* in such a way as to exert an impulse against all the

The Construction of a System

spheres *CCC* which touch it, it transmits to them the whole of its movement, and remains after that motionless like the sphere *B*. And without supposing that the ethereal particles are of spherical form (for I see indeed no need to suppose them so) one may well understand that this property of communicating an impulse does not fail to contribute to the aforesaid propagation of movement.[87]

Further,

Although the particles are supposed to be in continual movement (for there are many reasons for this), the successive propagation of the waves can not be hindered by this, because the propagation consists nowise in the transport of those particles but merely in a small agitation which they cannot help communicating to those surroundings, notwithstanding any movement which may act on them causing them to be changing positions amongst themselves.[88]

Huygens devised a complete scheme for the propagation and non-confusion of light transmitted in the ether. It was precisely this scheme that Newton had found unsatisfactory when he considered the subject.

Newton rejected Huygens's application of spherical wave fronts as an explanation of the transmission of light rectilinearly through a fluid. He did, however, agree with parts of Huygens's work. In Book II, Section VII of the *Principia* he gave Proposition XLI, Theorem XXXII, "A pressure is not propagated through a fluid in rectilinear directions except where the particles of the fluid lie in a right line." This proposition was in complete agreement with Huygens. The pressure would, in fact, be propagated to particles which are off the right line and the pressure would be propagated obliquely in the form of a spherical surface. In Proposition XLII, Theorem XXXIII, Newton wrote: "All motion propagated through a fluid diverges from a rectilinear progress into the unmoved spaces." Huygens would also have agreed; these principles apparently represented well-known properties. But now, in the Scholium, Newton concluded, "The last Propositions by respect the motions of light and sounds; for since light is propagated in right lines, it is certain that it cannot consist in an action alone (by Prop. XLI and XLII). As to sounds, since they arise from tremulous bodies, they can be nothing else but pulses of the air propagated through it (by Prop. XLIII).[89]

From a basic agreement on the principles of motion through a fluid, the two men drew different conclusions. Huygens found them satisfactory for light, Newton did not. These different conclusions seem to have been the result of two different points of view concerning light. Throughout his optical writings, Newton presented light in terms of light rays or light beams. He did not offer the interpretation of light as a wave or a pulse, although his suggestion of "Fits" involved the accompaniment of the light ray by a pulse or disturbance in the ether. Furthermore, Newton was concerned with the similarities between light and sound; he did not discuss the differences between them as Huygens had done. Newton argued that the explanation of the transmission of sound in air was considered to be identical to any attempted undulatory explanation of light in an ether. Newton's rejection of Huygens's arguments for the transmission of light was therefore based on the analogy between light and sound. Because he knew that sound diverged in air, he assumed that light would also diverge if it were an undulation. Because Newton observed light rays to be transmitted rectilinearly in a way similar to small bodies, he concluded that light could not be an undulation.

The objection of Newton and the Newtonians, that light could not be undulatory because, if it were, it would act like sound, really missed the point as a criticism of Huygens's own theory. As we have seen, Huygens's system specifically dealt with the dissimilarities. He did not maintain the oversimplified position that Newton took, but reserved for light a more subtle position, distinct from the simple analogy to sound. Huygens's conception of the combination of spherical wave fronts allowed him to distinguish light from sound. These spherical wavelets enabled Huygens to avoid a criticism similar to Newton's. Both Newton and the Newtonians failed to appreciate Huygens's arguments because they proceeded in optics as if light were a corpuscle, acted upon by forces. Newton's representation of the colors of the spectrum as distinct circles of color in the *Opticks* and his proofs of reflection and refraction of small bodies in the *Principia* were only the most obvious examples of Newton's apparent commitment to the consideration of light in terms of corpuscles. In short, both Newton and the Newtonians never gave Huygens's theory of light full consideration, because of the differences in their orientations.

The concept of light as a spherical wave had obvious potential, because Huygens's concentration on spherical waves resulted in the concept of the combination of spherical wave fronts. He devised this concept from a consideration of the various phenomena which indicated that light waves do not interfere with one another when crossing. Speaking of the non-confusion of the spherical waves, Huygens offered the following explanation as to why there was no confusion of the waves on crossing (see Fig. 7):

After all, this prodigious quantity of waves which traverse one another without confusion and without effacing one another must not be deemed inconceivable; it being certain that one and the same particle of matter can serve for many waves coming from different sides or even from contrary directions, not only if it is struck

The Construction of a System

Fig. 7 *Non-confusion of spherical waves.*

by blows which follow one another closely, but even for those which act on it at the same instant. It can do so because the spreading of the movement is successive. This may be proved by the row of equal spheres of hard matter, spoken of above. If against this row there are pushed from two opposite sides at the same time two similar spheres A and D, one will see each of them rebound with the same velocity which it had in striking, yet the whole row will remain in its place, although the movement has passed along its whole length twice over. And if these contrary movements happen to meet one another at the middle sphere, B, or at some other such as C, that sphere will yield and act as a spring at both sides, and so will serve at the same instant to transmit these two movements.[90]

It was after this introductory explanation that Huygens made one of his most important additions to the undulatory theory. His concentration on spherical waves became clear when he argued that not only does the particle "communicate its motion only to the next particle which is in the straight line drawn from the luminous point," but it also communicated its motion to all the other particles which touched it. Each particle was, therefore, the center of a spherical wave. These waves could be extremely feeble in themselves, but each wave contributed to the formation of a wave front which was of a magnitude dependent on the strengths of the individual contributing spherical waves.[91]

Huygens claimed the priority for this discovery for himself. He wrote:

And all this ought not to seem fraught with too much minuteness or subtlety, since we shall see in the sequel that all the properties of light, and everything pertaining to its reflexion and its refraction, can be explained in principle by this means. This is a matter which has been quite unknown to those who hitherto have begun to consider the waves of light, amongst whom are Mr. Hooke in his Micrographia, and Father Pardies, who in a treatise of which he let me see a portion, and which he was unable to complete as he died shortly afterward, had undertaken to prove by these waves the effects of reflection and refraction. But the chief foundation, which consists in the remark I have just made, was lacking in his demonstrations; for the rest, he had opinions very different from mine as may be will appear some day if his writing has been preserved.[92]

Huygens's contribution of the principle of a combination of wave fronts to produce a single wave front was indeed an important one, but it remained unappreciated in the eighteenth century. He made good his promise of describing reflection and refraction in terms of the spherical waves. He also provided an explanation for the ordinary and extraordinary rays produced by double refraction in Iceland spar.[93]

Huygens's analysis of optics in the *Treatise* was rather complete and it offered an accurate mathematical treatment of the major optical phenomena. The one great omission, however, was a treatment of color phenomena. This, of course, was one of the strongest parts of Newton's optics. Huygens did not want to include a treatment of colors in his work because he did not have an explanation for colors. He wanted a definite mechanical explanation for colors before he would include them. This was, to be sure, exactly the kind of explanation which Newton claimed not to have and not to have concerned himself with in his experimental philosophy.

Huygens also failed to press his concept of spherical wavelets to include the phenomena of diffraction. Grimaldi's experiments with diffraction phenomena were certainly known. It has been suggested that Huygens's neglect of diffraction "was probably due to a lack of the mathematical knowledge required to deal with the kinematics and dynamics of periodic motions."[94] Perhaps also, Huygens did not want to consider the bending of light into shadows because he was so concerned with demonstrating the rectilinear propagation of light in the ether. To begin to treat the case of diffraction after a demonstration of rectilinear propagation might have diminished the strength of arguments which were already criticized by Newton.

Newtonians such as Edmund Halley were quick enough in their criticism of Huygens's *Treatise*. A special point of controversy was the velocity of light. Halley wondered why the velocity of light was not uniform in all substances if light consisted of a motion in the ether. Since the ether permeated all bodies, why did the velocity of light not remain constant? Also, if the velocity of light were slower in denser media and faster in rarer, as Huygens supposed, what made the light accelerate again upon passing from a dense medium into a rarer one? How, too, could any body be opaque if ether existed everywhere, even in the most opaque bodies? Huygens attempted to devise mechanical explanations to meet these objections, but they did not satisfy the Newtonians, nor was Huygens himself fully pleased with them.

After Newton's publication of the *Opticks*, and after Cotes's arguments against the ether in his Preface to the *Principia*, the undulatory theory suffered great loss of prestige in England. In many cases, especially

The Construction of a System

those concerning the more common optical phenomena such as reflection, the explanation offered by the undulatory theory was hard to visualize. While an explanation was offered, it was complicated and it involved the ether. Contemporaries found it hard to conceive of tiny waves striking relatively rough surfaces and still reflecting perfectly according to the reflection law, with no noticeable confusion. Newton attacked this very point in his *Opticks*. In dealing with the problem of reflection from a glass surface, he claimed that no matter how fine the grains used to polish the surface, there would still be scratches and grooves. He wrote:

And therefore if Light were reflected by impinging upon the solid parts of the glass, it could be scattered as much by the most polished glass as by the roughest. So then it remains a problem how glass polished by fretting substances can reflect Light so regularly as it does. And this problem is scarce otherwise to be solved, than by saying that the Reflection of the Ray is effected, not by a single point of the reflecting body, but by some power of the Body which is evenly diffused all over its Surface, and by which it acts upon the Ray without immediate contact. For that the parts of Bodies to act upon Light at a distance shall be shown hereafter.[95]

Once again, Newton's criticism did not do Huygens justice. Huygens had shown clearly in his *Treatise* how this reflection could take place in terms of spherical wavelets in the ether. His treatment was exact and complete. The problem did not have to be "scarce otherwise solved" than by postulating an evenly diffused power.[96]

In general, in England, the undulatory theory received little support and widespread criticism. Newton's major objections to the theory, the analogy between light and sound, and his explanation of reflection were considered to be insurmountable by his followers. The Newtonians, with their preference for forces and particles of light, were unfair or at least narrow-minded in their consideration of the undulatory theory. The ether as Descartes and Huygens used it had little place in their scheme of things. As we have seen, the Newtonians had devised a system of optics which excluded such concepts as the ether and light as a wave.

While the undulatory theory of light was generally looked upon with disfavor in England, and especially so at the Royal Society,[97] it was not without adherents on the Continent during the eighteenth century. This was true for a variety of reasons, not the least of which was the Continental criticism which Newton received after the publication of the *Principia*. The famous Leibnitz-Clarke debate revealed the range and depth of this disagreement in science, religion, and philosophy.[98] There was general concern that Newton had re-introduced "occult" qualities into physics with his use of such words as "attraction" and "force."[99] Their allegiance to mechanical explanation made it difficult for Continental philosophers to accept the procedure followed by the English Newtonians of accepting attraction because it seemed to be based on observation and experiment. "Attraction" to the Newtonians was just a term for what was observed—no need to worry unduly about the cause of attraction when the application of the concept brought immediate results.[100] Cartesians considered Newton's theory of attraction a surrender to Aristotelian scholasticism and to the notions of immaterial sympathies and influences. To those who were disturbed by this throwback, there could be no comfort in the fact that the *Principia* and the *Opticks* did not depend upon these references to immaterial influences, although they did contain them. This is a good example of the continued influence of the past upon science just after Newton. The younger generation, such as Cotes in England and Voltaire in France, were not so caught by the remembrance of the old physics and had much less difficulty than the older men in accepting "attraction" and either forgetting about, or living with, references to ultimate causes. The eighteenth century was very similar to the twentieth in this respect; the new science was accepted after the old generation passed away.[101]

The work of supporters of the undulatory theory on the Continent reached England with regularity in the eighteenth century, but it was generally disregarded. One of the most important and most widely known of these continental theorists was Leonhard Euler.

Euler's preference for an undulatory theory of light can be related to his preference for physical explanation in terms of an ether. Euler used the conception of an ether continuously filling all space to explain the phenomena of light, heat, electricity, magnetism, and even gravitation. With this orientation, the undulatory theory which utilized the ether was certainly more convenient than a system using forces and particles. Euler's arguments in support of an undulatory theory are of interest to us here because they represent a rather good summary of eighteenth century objections raised toward the Newtonian theory of light.

Euler raised most of the objections to the Newtonian system that were available in the eighteenth century. His arguments were not very forceful, however, and his overextension of simple analogies was very similar to the overemphasis placed by the Newtonians on the analogy of light and sound. Here again, we have an example of the limitations imposed by a definite orientation toward one explanation of the physical world. Euler's arguments by analogy were only satisfactory to someone who already believed in the basic premises involved. For example, Euler compared the streams of light particles which were supposed to compose a ray of light in the Newtonian system to the water particles composing the stream of a fountain. Euler argued that all luminous bodies should therefore experience a weight loss. Since the particles of light were particles, no matter how subtle they were, they were emitted in such great quantity from a luminous body that the body must eventually lose weight. This type of argument was persuasive enough from the point of view of a person who thought that an undulation in the ether, without any possible weight loss, was a more acceptable explanation. However, from the point of view of a Newtonian, believing in action-at-a-distance and a void, since there must be "something" to be acted upon, light must consist of particles so subtle and so small that they did not in fact produce a detectable weight loss. It was as easy for a Newtonian to maintain that the particles of light were extremely subtle as it was for Euler to believe in an all-pervasive ether. Arguments alone were not effective in establishing the superiority of either system since both had problems associated with them. Without a new experimental or decisive theoretical discovery, neither system could be defended definitively. The choice between them remained largely a matter of personal preference during the eighteenth century. Generally, the English preferred to use concepts derived from the *Principia* and *Opticks*. The Newtonian system of optics was clearly more compatible with this preference.

Euler did make important contributions to the undulatory theory, even though his arguments did not win wide support. He took up the explanation of colors which Huygens had avoided, using a mechanical explanation suggested by Malebranche. Euler stressed the parallel between the pitch of sound and the color of light. He established the concept, within the undulatory framework, that the frequency of vibration of a musical chord had a parallel in the number of vibrations which constituted colors. This concept was not new, but Euler stated it clearly and concisely as part of the undulatory theory.

Malebranche had proposed a modification of Descartes's theory of colors which was based on the concept of vibration. Malebranche advanced the idea that colors, rather than being produced by the spin of ether globulei as Descartes suggested, were instead produced by a vibration of tiny ether vortices which he thought composed light. Malebranche substituted ether vortices for globulei and offered his explanation of color phenomena by describing the effect of different conditions on the vibration of these vortices. The vortices were too cumbersome and strange a concept to find acceptance, especially after the second edition of the *Principia*. The idea of the relation of colors to frequency was, however, brought forward by Malebranche.[102]

Euler put a great deal of reliance upon the concept of frequency of vibration in relation to colors. He developed this concept in two published works, in the *Memoirs of the Academy of Berlin*[103] and in his *Letters to a German Princess*.[104] The relationship was stated most clearly in his letter XXVII to the Princess of Anhalt-Dessay.

> The parallel between sound and light is so perfect that it holds even in the minutest circumstances. When I produced the phenomenon of a musical chord, which may be excited into vibration by the resonance only of certain sounds, you will please to recollect, that the one which gives the unison of the chord in question is the most proper to shake it, and that other sounds affect it only in proportion as they are in consonance with it. It is exactly the same as to light and colours; for the different colours correspond to the different musical sounds.
>
> The colours are produced then by specific vibrations . . .
>
> And the rays which make such a number of vibrations in a second may, with equal propriety, be denominated red rays; . . .[105]

Euler did not develop this suggestion in any systematic way. It was just a part of his undulatory scheme of light. This scheme had none of the coherence or completeness of the Newtonian system. More important, Euler did not make use of Huygens's ideas of spherical wavelets. The undulatory theory of light remained disjointed, with only haphazard support during the eighteenth century.

There certainly were attempts to include light in the scheme of imponderable fluids during the eighteenth century, especially after 1740.[106] But these attempts could not seriously challenge Newtonian optics because they were all concerned with broader topics of chemistry, or electricity, or phlogiston. Light was always presented as something which should be included in the new materialist framework. The arguments in favor of a re-interpretation of light in conformity with imponderable fluids were always of a general nature, based upon assertions about the nature of matter. Being general, and not offering any experimental evidence contradicting the Newtonian system, they simply formed part of the background for considerations about the nature of light. Since there was little agreement about imponderable fluids in general, and since light was only one of the imponderables to be considered, nothing very forceful emerged from the attempts to bring light into the scheme of imponderable fluids. The work of Brian Higgins is a good case in point.

Brian Higgins wrote *A Philosophical Essay Concerning Light* in 1776.[107]

The Construction of a System

Higgins began with the attempt to include light in the framework of materialistic explanations in chemistry. He was especially interested in relating light and phlogiston, and in entertaining arguments for the possiblity "that the electrical fluid consists chiefly of Light." He convinced himself by a number of qualitative arguments that light could not have a "progressive motion" nor was light "projected or moved in the manner described in the *Newtonian* schools."[108] He relied heavily upon the use of analogy, as illustrated in the following paragraph:

> If we consider the Light contained in illuminated spaces, we cannot from any analogy presume that illumination is continued by any progressive motion of Light. When a stone is thrown into stagnant water, the motion of the parts of water contiguous to the stone, is communicated to the circumambient water; and the extreme circular waves which reach the banks, do not consist of the parts first impelled, but of the water which was contiguous to the banks; and if the stone be small relatively to the water, the motion of the water is manifestly undulatory, where it is not sensibly progressive. A person immersed in water, can hear the collision of two stones in the water, at a great distance, as the Learned Dr. *Franklin* observes. In this case the whole water hath no progressive motion, altho' motion is propagated in it to the greatest distances at which the experiment has been made.[109]

Higgins continued this analogy of the motion of sound. He was convinced that light, as a corpuscle, acted upon by forces, was not compatible with motions of chemistry, electricity, and phlogiston. In this, of course, he was quite right. Corpuscular mechanics proved inadequate to meet the demands of the vast number of new experiments in these fields. Reasoning by analogy, Higgins tried to show that the corpuscular interpretation of light also had deficiencies in this materialist framework. He turned back to Huygens and Euler for support:

> ... But I must observe here, that Huygens and others who wrote before *Newton* entertained a different opinion; and that M. *Le Cat*, M. *Euler*, and many distinguished modern philosophers agree in rejecting this part of the *Newtonian* theory of Light; and that M. *Euler* especially has lately made some advances toward explaining the phenomena, by motion propagated in a subtil medium.[110]

This passage reveals the problem faced by those who wished to include light in a materialist framework. The "part of the *Newtonian* theory of Light," the progressive motion of the corpuscles of light, simply could not be rejected without the whole structure of Newtonian optics collapsing. This, of itself, would have been acceptable, had there been an alternative. But, as Higgins tactfully indicated, M. Euler had only made "some advances" in that direction. The corpuscular theory of light was not compatable with work in chemistry, electricity, and phlogiston at the end of the eighteenth century. But it was compatible with mechanics and astronomy, as we have seen. Only through the work of Thomas Young in the opening years of the nineteenth century did the undulatory theory of light assume a unified form, effective enough to seriously challenge the Newtonian system of optics. Previous objections had been voiced, but none were based upon new experimental and theoretical work, as were Young's. His undulatory theory posed the first serious challenge to Newtonian optics in England. The significance of Young's work can now be treated, with the context of eighteenth century optics in mind.

III
Thomas Young and the Challenge to Newtonian Optics

Thomas Young's Concept of Interference

An undulatory theory of light was advanced in England by Thomas Young during the opening years of the nineteenth century. The undulatory theory had found little favor in England during the preceding century and Young's work in optics was at first rejected by his English contemporaries in favor of the traditional Newtonian system of optics. Young's work stood apart from the Newtonian tradition, and his interest in the undulatory conception of light represented a deviation from the accepted English point of view. Young found the interest and inspiration for his work on light in sources rather different from those of his Newtonian predecessors. He did not use Newton's *Opticks* as his base, and his work was not a continuation of thoughts suggested by Newton. It is the purpose of this chapter to show how Young's work developed outside the Newtonian tradition which we have considered. Young's introduction to the study of light occurred in a way quite different from any suggestions in Newton's *Opticks*.[1] The steps in Young's break with the Newtonian tradition will be considered after a study of his early education and work.

Young's biographers gave a clear picture of his early childhood and interests. He was an exceptionally gifted child with a capacity for learning languages, and a keen, natural curiosity for all new areas of learning. At about the age of twelve, Young became interested in botany. He decided to build a microscope to use for the closer study of the plants he collected. A contemporary wrote, "he attempted the construction of a microscope from the descriptions of Benjamin Martin. This led him to Optics . . . ."[2] He was also interested in the study of fluxions at this time, although apparently learning nothing of the Continental developments in mathematics. Young's early years were spent following a self-taught path to a wide range of accomplishments.

In 1792, at the age of nineteen, he began his medical education. The choice of the medical profession seemed to be, in part, Young's natural inclination and, in part, the advice of several doctors and family friends.[3] Young entered wholeheartedly into the study of medicine, studying at

London, Edinburgh, Göttingen, and Cambridge. His primary goal in this full career of study was the attainment of a well-balanced, full education to best prepare himself to become a good physician. Medicine was Young's major concern; however, his secondary interests ranged widely during his student years. These interests were directed first toward a study of vision and then, in Germany, to a study of sound. When he returned to England to complete his medical preparations, Young found the leisure to reconsider and combine his interests in light and sound. It was the combination of various thoughts on both subjects which led Young to the formulation of his undulatory theory of light. The steps in this formulation will now be considered in detail; they show Young approaching the theory through his medical interests and not through an extension of hints provided by Newton's *Opticks*.

Young's first interest in light was derived mainly from his medical interest in the study of human vision. His first medical paper, entitled "Observations on Vision" was read to the Royal Society on May 30, 1793. Young recorded in his personal bibliography:

4. Observations on Vision: Philosophical Transactions 1793, p. 169, explaining the accommodation of the Eye, from a muscular power in the crystalline lens—but immediately afterwards claimed by John Hunter, as a discovery of his own.[4]

His work on the eye was motivated by an interest in the physical structure of the human eye. It required an understanding of the fundamentals of geometrical optics to determine the relationship between focal points and lens proportions. There was no indication of any further theoretical interest, but the paper did display an understanding of geometrical optics. It was on the basis of this paper, as well as the influence of Young's numerous friends, that he was elected a Fellow of the Royal Society on June 19, 1794. He became widely known at this time as a bright, accomplished young man.

Young became interested in sound through his study of the human voice. This research on the voice was a medical interest, undertaken during his period of study at Göttingen. His doctoral dissertation at Göttingen in 1796 was entitled, "De Corporis Humani Viribus Conservatricibus." This thesis seems not to have survived, but a fragment

gives an alphabet of forty-seven letters designed to express, by their combination, every sound which the organs of the human voice are capable of forming . . . .[5]

Young returned to England after his dissertation, with every intention of taking up a medical practice. But, because of a complication in the laws governing medical practice in London, he was forced to attend six terms at Cambridge in order to obtain a degree which would allow him to practice in London. In a letter to his friend Dalzel in Edinburgh, Young wrote:

> The foolish laws of the college in London are perplexed and ill-understood; but I must now make the best of Cambridge.[6]

From the point of view of the history of science, Young certainly did make the best of his Cambridge days. He was admitted to Emmanuel College, Cambridge, as a Fellow Commoner on March 18, 1797. Being subject to few of the disciplinary rules for regular students, Young was free to pursue his own interests. He certainly needed no further medical training so he was free to continue his interests in science. In the summer of 1797, Young wrote to Dalzel from Cambridge:

> I am at present a good deal employed on the subject of the light synoptic sketch at the end of my thesis, the definition and classification of the various sounds of all the languages that I can gain knowledge of; and have of late been diverging a little into the physical and mathematical theory of sound in general. I fancy I have made some singular observations on vibrating strings, and I mean to pursue my experiments.[7]

His study of voice led him naturally to a study of vibrating strings and also to a study of air flow. His interests gradually moved from purely medical concerns to studies of a more abstract nature. In the early summer of 1798, Young gave the first evidence of detailed theoretical speculation and the first use of Continental science. In a letter to his friend, Dr. Bostock, he wrote:

> ... I have been studying not the theory of winds, but of the air, and I have made observations on harmonics which I believe are new. Several circumstances unknown to the English mathematicians which I thought I had first discovered, I since find to have been discovered and demonstrated by foreign mathematicians; in fact Britain is very much behind its neighbors in many branches of the mathematics; were I to apply deeply to them I would become a disciple of the French and German school; but the field is too wide and too barren for me.[8]

Young made rapid progress in the study of sound, harmonics, and the mathematics of vibrating strings. He had a special interest in music and started reading widely in the field of musical theory, including Robert Smith's *Harmonics*. In July of 1798, Young summarized his attainments in a letter to Dalzel:

> I have been lately pursuing a little further the theory of sound, and among other papers have read that which you mentioned to me of Mr. Walter Young of rhythm.

He had treated the subject in a very masterly manner, and his dissertation is well worthy the attention of critics as well as musicians. I am ashamed to find how much the foreign mathematicians for these forty years have surpassed the English in the higher branches of sciences. Euler, Bernoulli, and D'Alembert have given solutions of problems which have scarcely occured to us in this country. I have had particular occasion to observe this in considering the figure of vibrating chords, the sounds of musical pipes and some other similar matters in which I fancied I had hit on some ideas entirely new, but I was glad to find them in part anticipated by Bernoulli in 1753 and 1762. There are still several particulars respecting the gyration of chords, the formation of synchronous harmonies, the combination of sounds in air, the phenomena of beats, on which I flatter myself that I shall be able to throw some new light, and to correct several mistatements of Dr. Smith, whose work I think has only been admired because few would trouble themselves to wade through so much affected obscurity.[9]

We see here not only the level of Young's attainment, but also his attitude toward the state of English science. Moreover, he took the point of view of someone fully capable of understanding the Continental developments in mathematics. He was anxious to perform his own theoretical investigation and would not allow the "old" accepted works of English science to remain unchallenged. Young was the first English natural philosopher in the nineteenth century to actively pursue a line of investigation based on an understanding of Continental science and mathematics and consciously critical of the English tradition.

In the spring of 1799, Young "kept" the six terms necessary for his degree and moved to London to begin the profession for which he had prepared so long and thoroughly. The scientific results of his two years of relative leisure at Emmanuel College became apparent when Young published a paper entitled "Outlines of Experiments and Inquiries Respecting Sound and Light," dated Emmanuel College, Cambridge, 8th July, 1799.[10] Young made his thoughts on light and sound public in this paper. Evidently he had done a great deal of thinking and studying on these subjects. He also made it clear that these topics were not simply of passing interest to him, but would be subject to his continued interest and experiment.

It has long been my intention to lay before the Royal Society a few observations on the subject of sound; and I have endeavored to collect as much information and to make as many experiments connected with this inquiry as circumstances enabled me to do; but the further I have proceeded the more widely the prospects of what lay before me has been extended, and . . . I find that the investigation, in all its magnitude, will occupy the leisure hours of some years, or perhaps a life . . . .[11]

Young was dissatisfied with the Newtonian system of optics, and his two years at Cambridge had convinced him that this system was unsatisfactory and should be replaced. The purpose of his first paper on light and sound was to provide a basis for the reassertion of the undulatory theory in England. Young wanted to replace the Newtonian system with a system which he had found more correct. His arguments proceeded generally on three grounds: (1) that Newton's rejection of the undulatory theory of light because of the supposed impossibility of rectilinear propagation of undulations was mistaken, (2) that the corpuscular system had certain unanswerable difficulties previously overlooked, and (3) that the comparison between phenomena of sound and phenomena of light could be greatly strengthened by a new analogy.

Young devoted the first quarter of his paper to the detailed account of various new experiments that he had performed while at Cambridge, experiments on the divergence of streams of air flowing through apertures and on the "ocular evidence for the nature of sound" obtained by the method of blowing smoke through tubes producing notes. The reason for all these experiments suddenly became clear in Section VI of the paper, "Of the Divergences of Sound."

It has been generally asserted [Young wrote] chiefly on the authority of Newton that if any sound be admitted through an aperture into a chamber, it will diverge from that aperture equally in all directions. The chief arguments in favour of this opinion are deduced from considering the phenomena of the pressure of fluids, and the motion of waves excited in a pool of water. But the inference seems to be too hastily drawn: there is a very material difference between impulse and pressure . . . .[12]

Young noted that Newton had based his rejection of the undulatory theory of light on the mistaken conclusion about the propagation of sound and his too literal acceptance of the exact analogy between light in the ether and sound in air. Newton's mistake occurred when he assumed that the transmission of an impulse through a particulate fluid was analogous to the propagation of sound through air. This mistake had been accepted throughout the eighteenth century. Young argued against this misconception with the support of his own new experiments.

From the experiments on the motion of a current of air, already related, it would be expected that a sound, admitted at a considerable distance from its origin through an aperture would proceed, with an almost imperceptable increase of divergence, in the same direction; for, the actual velocity of the particles of air, in the strongest sound, in incomperably less than that of the slowest of the currents in the experiments related, where the beginning of the initial divergence took place at greatest distance.[13]

Young actually set up experiments to test Newton's line of argument. By an arrangement of tubes and streams of smoke, he was able to check visually the amount of divergence occurring at differing velocities. The apparatus was quite simple, but cleverly conceived. Young was able to fully convince himself that Newton's objections to the rectilinear propagation of undulations were a priori and mistaken.

After overcoming this major objection to an undulatory theory, Young continued by criticizing the Newtonian theory of light. His approach was moderate, but definitely sceptical. A section from Young's work is worth quoting at length here, because it gives a clear example of his style of reasoning and it illustrates his relation to the undulatory theorists of the past:

Ever since the publication of Sir Isaac Newton's incomparable writings, his doctrines of the emanation of particles of light from lucid substances, and of the formal pre-existence of coloured rays in white light, have been almost universally admitted in this country, and but little opposed in others. Leonhard Euler indeed, in several of his works, has advanced some powerful objections against them, but not sufficiently powerful to justify the dogmatical reprobation with which he treats them; and he has left that system of the ethereal vibration, which after Huygens and some others he adopted, equally liable to be attacked on many weak sides. Without pretending to decide positively on the controversy, it is conceived that some considerations may be brought forwards, which may tend to diminish the weight of objections to a theory similar to the Huygenian. There are also one or two difficulties in the Newtonian system which have been little observed. The first is, the uniform velocity with which light is supposed to be projected from all luminous bodies, in consequence of heat or otherwise. How happens it that, whether the projecting force is the slightest transmission of electricity, the friction of two pebbles, the lowest degree of visible ignition, the white heat of a wind furnace, or the intense heat of the sun itself, these wonderful corpuscles are always propelled with one uniform velocity? For, if they differed in velocity, that difference ought to produce a different refraction. But a still more insurmountable difficulty seems to occur, in the partial reflection from every refracting surface. Why, of the same kind of rays, in every circumstance precisely similar, some should always be reflected and others transmitted appears in this system to be wholly inexplicable. That medium resembling, in many properties, that which has been dominated ether, does really exist, is undeniably proved by the phaenomena of electricity; and the arguments against the existence of such an ether throughout the universe, have been pretty sufficiently answered by Euler. The rapid transmission of the electrical shock, shows that the electric medium is possessed of an elasticity as great as is necessary to be supposed for the propagation of light.

Whether the electric ether is to be considered as the same with the luminous ether, if such a fluid exists, may perhaps at some future time be discovered by experiment; hitherto I have not been able to observe that the refractive power of a fluid undergoes any change by electricity. The uniformity of the motion of light in the same medium, which is a difficulty in the Newtonian theory favours the admission of the Huygenian; as all impressions are known to be transmitted through an elastic fluid with the same velocity. It has been already shown, that sound, in all probability, has very little tendency to diverge: in a medium so highly elastic as the luminous ether must be supposed to be, the tendency to diverge may be considered as infinitely small, and the grand objection to the system of vibration will be removed.[14]

Although Young argued for the acceptability of the undulatory theory, he realized that there were several unexplained problems. He was not, at this point, convinced of the superiority of the system in all its aspects. He had merely performed some experiments on sound and, after finding Newton's objection in error, had become favorably disposed toward the Huygenian system. Both the argument against the possibility of uniform velocity for all corpuscles of light and the failure to account for partial reflection and refraction swayed Young toward a system similar to that proposed by Huygens. Young was convinced that the existence of an ether was possible because of his familiarity with work concerning electric fluids, especially that of Euler. He was also impressed by the possibility of explaining the uniform velocity of light in terms of the uniform velocity of waves in the ether. But he was not convinced by Euler's argument that the ether, which served to carry the undulations of light, was the same ether which produced electrical and even gravitational effects. Also, he was not satisfied that Euler's explanation of the refraction process was without difficulty. He did, however, temporarily accept an explanation of refraction very similar to Euler's in terms of ether density, because he had no alternate suggestion to make.[15]

Young quickly passed through the problems involved with explanations based on the undulatory system and the ether. He was not prepared to support the theory in all its aspects. Rather, he wished to present those arguments which he found favorable to the system, leaving a more complete treatment for the future. He moved quickly from the problems of the undulatory system, which, as he wrote, remain "to be hereafter determined," to the important comparison of the colors of thin plates and the sounds in a series of organ pipes.

Young used the analogy of thin plates and organ pipes to support Euler's suggestion "that the colors of light consist in the different frequency of the vibrations of the luminous ether." Euler had merely suggested this relation without providing any support for its applicability. Young found support for the relation by using both his new law of interference and the careful experiments of Newton. He rejected Newton's explanation of the colors of thin films using "Fits of Easy Transmission and Reflection." He did accept, however, Newton's account of the recurrence of colors corresponding to a given recurrence of thickness in arithmetic progression. Young made use of Euler's suggestion and Newton's description as follows:

Now this is precisely similar to the production of the same sound, by means of a uniform blast, from organ pipes which are different multiples of the same length. Supposing white light to be a continued impulse or stream of luminous ether, it may be conceived to act on the plates as a blast of air does on the organ pipes, and to produce vibrations regulated in frequency by the length of lines which are terminated by the two refracting surfaces.[16]

Young innocently used this analogy to support the suggestion that color phenomena on thin films were similar to vibrations of sound in organ pipes. He did not, at this time, propose any process of interference operating in the phenomena of light and colors. However, he did realize that the concept of interference of undulations could be applied to the production of sound in organ pipes. In Section xi "Of the Coalescence of Musical Sounds," Young established the principle of interference in sound relating to the production of beats and grave harmonics. Light was not included in this first statement of the principle of interference.

Young's realization of the principle of interference stemmed from a critical analysis of Robert Smith's *Harmonics*. His interest in music and sound had led him to read Smith's *Harmonics*, an accepted and respected textbook on the subject. In Smith's work, Young came upon an incorrect description of the crossing of sound waves. Realization of the inaccuracy of this description played a major role in stimulating Young to formulate the theory of interference. In his own words:

It is surprising that so great a mathematician as Dr. Smith could have entertained for a moment, an idea that the vibrations constituting different sounds should be able to cross each other in all directions without affecting the same individual particles of air by their joint forces: undoubtedly they cross, without disturbing each other's progress; but this can be no otherwise effected than by each particle partaking of both motions. If this assertion stood in need of any proof, it might be amply furnished by the phaenomena of beats, and of the grave harmonics observed by Romien and Tartini; which M. DeLaGrange has already considered in the same point of view. In the first place to simplify the statement, let us suppose what probably never precisely happens, that the particles of air, in transmitting the pulses, proceed and return with uniform motions; . . . let the uniform progress of time be represented by the increase of the abscissa, and the distance of the particles from its original position, by the ordinate . . . . Then, by supposing any two or more vibrations in the same direction to be combined, the joint motion will be represented by the sum of the difference of the ordinates. When two sounds are of equal strength, and nearly of the same pitch . . . the going vibration is alternately very weak and very strong producing the effect denominated a beat . . . which is slower and more marked, as the sounds approach nearer to each other in frequency of vibrations . . . . The strength of the joint sound is double that of the simple sound only at the middle of the beat, but not throughout its duration and it may be inferred, that the strength of sound in concert will not be in exact proportion to the number of instruments composing it. Could any method be devised for ascertaining this by experiment, it would assist in the comparison of sound with light.[17]

Young supported his argument using beats by describing the phenomenon of the production of a "grave harmonic." This "grave harmonic" had no relation to the "acute harmonic"—the harmonic produced by the vibration of a single string. The "grave harmonic" was one sound produced by the interference of the sound from two strings, representing, therefore, a compound vibration. It was the production of a new tone by the coalescence of two other tones. When two strings, each of nearly the same frequency, are sounded, a beat is heard which has a frequency corresponding to the difference in the frequencies of the two strings. But as the interval between the two strings becomes widened, the frequency of the beats becomes too great for the individual beat to be distinguished. In this case, a tone is heard, which like the beat has a frequency equal to the difference in frequency of the two strings. This combination-tone is called the grave harmonic. The grave harmonic was quite familiar to musicians of the eighteenth century. It was impossible to explain on the basis of Smith's *Harmonics*. Young offered an explanation in terms of interference.

Young now had all the elements of the theory of interference, but at the time of this paper, he did not realize its applicability to the phenomena of colors. There is enough evidence available, both in Young's own words and in the way in which he replied to his first critics, to indicate that he came to a full realization of the applicability of the law of interference to light and colors between the completion of his first paper July 8, 1799, and August, 1801, the date of his first reply to criticism in Nicholson's *Journal*.[18] The analogy between sound in organ pipes and color in thin plates needed only to be associated with his thoughts on the coalescence of sounds for him to formulate the law of interference. Young wrote to a friend:

I am at present employed in some further optical investigations [Young wrote to Dalzel in June, 1801] which, I imagine, will be considered as more important than any of my former attempts, as I think they will establish almost incontrovertibly the undulatory system of light, and extend it to the explanation of an immense variety of phenomena.[19]

Young's own words support the idea that the principle of interference originated first from speculations on the coalescence of sounds. After its realization, Young thought to apply it to the colors of thin films and light.

It was in May 1801 [Young wrote] that I discovered by reflecting on the beautiful experiments of Newton, a law of which appears to me to account for a greater variety of interesting phenomena than any other optical principle that has yet been made known . . . and this I call the general law of the interference of light, I have shown that this law agrees, most accurately, with the measures recorded in Newton's *Opticks* relative to the colours of transparent substances . . . there was nothing that could have led to it in any author with whom I am acquainted, except some imperfect hints in those inexhaustible but neglected mines of nascent inventions, the works of the great Dr. Hooke, which had never occurred to me at the time that I discovered the law, and except the Newtonian explanation of the combination of tides in the Port of Batsha.[20]

As Young stated, the law of interference might have arisen from speculation on Hooke's *Micrographia*. But there seems every reason to believe Young when he wrote:

It was not till I had satisfied myself respecting all these phenomena, that I found in Hooke's *Micrographia* a passage which might have led me earlier to a similar conclusion.[21]

The conception of interference might also have come from consideration of Halley's and Newton's considerations on the tides in the Port of Batsha in the East Indies.[22] The tides in this port were observed to occur in a strange manner. Instead of the usual two flood and two ebb tides per day, as occurred in ports on the Atlantic Ocean, the tides at Batsha rose and fell only once a day. In addition, there were two days per month which had very small tidal variations, sometimes none at all.[23] Newton attempted to explain the peculiar phenomena of the tides in Batsha in his proposition on tides. In his solution, Newton suggested that perhaps the flood and ebb tides which flowed toward the port, met on the days of small tidal activity. The meeting of the flood and ebb tides would have the effect of balancing one another. The result would be a stagnation of tidal movement for those days.[24] The solution Newton advanced suggested some type of combination of the tides to produce stagnation of tidal movement. The combination of the tides produced the observed quiet days of tidal activity.

It seems likely that Young recognized Newton's explanation of the tides as a particularly good example of the law of interference, after he had come to a full realization of the law on his own. Young's interest was concentrated on the phenomena of sound and his first expression of the principle of interference related to sound. As he stated, he discovered the applicability of the principle of interference to light in May, 1801. After Young realized that the concept of interference could be applied to light, he presented his law of the interference of light to the Royal Society in the Bakerian lecture on November 12, 1801.[25] In this important paper, no mention was made of the example of tides. That example was used, however, in a slightly modified form, using a lake and waves in a channel, in Young's *Reply* to Henry Brougham's criticisms in 1804. It is most probable that Young came upon the example of tides in preparation of his *Course of Lectures on Natural and Experimental Philosophy*, given in connection with his new position as Professor at the Royal Institution.

Young was appointed to the Royal Institution as Professor of Natural Philosophy on the recommendations of Count Rumford and Joseph Banks on August 3, 1801. He also took up joint responsibilities with Davy as editor of the *Journal* of the Royal Institution in the fall of 1801. Young must have been preparing his lectures in the fall and winter because he began his course on January 20, 1802, and finished on May 17, 1802. He also published a syllabus for the course on January 19, 1802. Young was not unaware of the possibility of his appointment as early as June 1801, but it seems unlikely that he would prepare his lectures before notification of his appointment.

Young's use of the example of the combination of tides appeared in his lecture on "Tides" in the third and last section of his lectures, headed "Part the Third—Physics." It was the second part, "Hydrodynamics," which contained his lectures on sound and light. It seems likely that Young recognized the phenomena of tides as a good example of interference sometime in December 1801, as he completed the last part of his course of lectures. Young did not use this example in November 1801, in his Bakerian Lecture, which, of course, was written sometime before this date. It seems probable that Young came to the realization of the law of interference not with the aid of thoughts on the tides but out of the context of his work on sound and his criticism of Robert Smith's *Harmonics*. As he himself tells us, after he developed the law for sound, he realized its applicability to light phenomena. Young continued his analogy between beats in organ pipes and the colors of thin films and formulated the general law of interference applied to light, working within the context of sound and light phenomena. As has been shown, the law of interference could emerge from just this research alone. While it is not impossible that the Newtonian work on combination of tides could have played a part, it seems outside the context of Young's work when he formulated the law of interference. It seems more probable that Young studied the tides as he prepared his last lectures on natural philosophy, after the Bakerian Lecture of 1801 was completed and the general law of interference had been stated.

Young's statements of the new theory of the coalescence of musical sounds and his realization of the applicability of the law of interference to colors, placed him squarely in fundamental conflict with the accepted Newtonian conceptions of light and sound. His presentation of the Bakerian lecture made it certain that his statements could not be ignored. The first critical comments on his work resulted from Young's theory of musical sounds. By specifically attacking Robert Smith's *Harmonics*, he had struck a direct blow at the traditional position with respect to sound. His first critics were most conservative in their counter-assertions and simply attempted to maintain the Newtonian tradition.

Professor Robison of Edinburgh was the first to publish a criticism of Young's work. He called Young to task for making disparaging remarks about Robert Smith's *Harmonics*. "We are surprised [sic] to see this work of Dr. Smith greatly undervalued," he wrote, "by a most ingenious gentleman in the philosophical transactions for 1800. . . ."[26] Young's criticism had been rather blunt:

Dr. Smith has written a large and obscure volume, which, for every purpose but for the use of an impracticable instrument, leaves the whole subject precisely where it found it.[27]

Young's statement had two bases. One was a dispute, of slight importance, over the system Smith had devised to obtain a more perfect harmony when the harpsichord was played in certain keys. The instrument required a special tuning which made it impractical for common use.

The second basis for criticism was the method of tuning organ pipes set forth by Robert Smith. Robison had made much of Smith's method in his article. Young disagreed. In Young's disagreement, we have further evidence of the important role his interest in music assumed in the formulation of the law of interference. His interest in the coalescence of musical sounds clearly arose from his criticism of Smith's work.

I do not mean to be understood that this work is so contemptible [Young wrote] as not to contain the least particle of important matter; but it appears to me that its errors counterbalance its merits. The only improvement on which Professor

Robison himself seems to set a high value, is the application of the phaenomena of beats to tuning an instrument: on the other hand, I conceive that the misstatement relative to the non-interference of different sounds, is an inaccuracy which far outweighs the merit of Dr. Smith's share of that improvement.[28]

Objections of a more specific nature to Young's theory of coalescence were raised by John Gough. Young and Gough exchanged several published letters on the theory of compound sounds.[29] Gough maintained that compound sounds were simple mixtures of sound waves, and were not, as Young maintained, formed by coalescence. This theory of simple mixture reflected Smith's views in the *Harmonics*.

I profess to maintain [Gough wrote] compound sounds to be mixtures of elementary sounds, not aggregates by coalescence; in other words, I undertake to defend Dr. Smith's position as a fundamental maxim both of harmonics and the general philosophy of sounds.[30]

Young answered Gough's comments quite civilly by clarifying the position he had taken in his 1799 paper. Gough's resistance stemmed primarily from a misunderstanding of the conditions necessary for coalescence; namely, the similarity of the direction necessary for coalescence and beats to occur. Young quoted Smith's work at length,[31] to show specifically which points of the old system were most difficult to maintain. After reiterating his statements in support of coalescence, he rested his whole argument on the example of grave harmonics. This argument was decisive, for it was inexplicable in terms of simple mixture:

I can not however avoid remarking [Young concluded] that Mr. Gough has wholly omitted to notice the fundamental fact which I stated as affording the most satisfactory of all proofs of the coalescence of sounds; that is, the production of a faint, but very audible graver sound from the union of two acute ones, a phenomena so well known to musicians, that I can scarcely suppose a person, who is ignorant of it, properly qualified to discuss the subject of harmonics. . . . Here is a product of combination which possesses properties totally different from those of its constituent parts; its pitch is always much lower, its quality of tone is perfectly singular.[32]

Young was convinced that his argument for coalescence was not only reasonable, but firmly rooted in experimental demonstration.

Young's first 1800 paper aroused some criticism from men who supported the traditional theory of sound. His Bakerian Lecture of November 1801, published in 1802, was destined to arouse a great deal more criticism from the adherents of Newtonian optics. The replies to his first paper had shown Young that he could expect criticism. In his

Bakerian lecture to the Royal Society, he decided to use an unusual technique of presentation. Rather than continuing to reassert his own reasoning and experiments in favor of an undulatory theory, Young sought to avoid criticism by making careful use of previously accepted authorities and known experiments of proven accuracy, attempting to give his law of interference the cloak of respectability by an appeal to authority. In his search for authorities, Young found, after a more complete reading of Newton's works, that he could effect a neat switch on the Newtonian tradition by ascribing the basic tenets of the undulatory theory to Sir Isaac Newton himself.

A more extensive examination of Newton's various writings has shown me, [Young wrote] that he was in reality the first that suggested a theory as I shall endeavor to maintain; that his own opinion varied less from the theory than is now almost universally supposed. . . .[33]

Young brought forth all the passages from Newton's works that the eighteenth century partisans of the corpuscular view had carefully neglected. The passages were out of context and were used to Young's advantage. In his lecture, Young quoted extensively from Newton, supporting with Newton's own words three concepts essential to the undulatory theory. Newton was made to appear to maintain Young's points, that: "(1) A luminiferous Ether pervades the universe, rare and elastic in a high degree, (2) Undulations are excited in the Ether whenever a body becomes luminous, and (3) the sensation of different colours depends on the different frequency of vibrations, excited by light in the retina."

It seems likely that Young attempted to use Newton as a support for his theory in order to put an end to objections by conservative Newtonians. Young had experienced public criticism of his theory of sound from men who sought to maintain the old, established Newtonian system. He must also have undergone private criticism of both the theory of sound and of interference applied to light and colors. His use of the quotations from Newton can be seen as a clever maneuver to avoid this type of purely conservative, non-constructive criticism. But more important, it can be considered as evidence of an astute perception that Newton's works were not as strictly consistent as his followers had maintained. This perception went to the very core of eighteenth century optics. Young called into question the whole system of optics which the various Newtonians, with their arguments in terms of forces and corpuscles, had attempted to establish. Rather than using Newton as a source of inspiration, Young used Newton as a means of challenging the position taken by the Newtonians during the eighteenth century. What better device could have been employed to demonstrate the inadequacy of the emission theory? Young used passages from the originator of the theory which contrasted with the Newtonian theory as it had been developed during the eighteenth century. He reversed Newton's position from the founder of the emission theory to a supporter of the undulatory theory, which put Young in an excellent debating position.

Young took a moderate approach, however, in his attempt to establish the undulatory theory. He knew that there were decisive arguments favoring many features of his theory, particularly the phenomena of the colors of thin films and striated surfaces, but there were also aspects which were more difficult to support. In arguing for the acceptance of the undulatory theory, Young wrote the following:

Although the invention of plausible hypotheses, independent of any connection with experimental observations, can be of very little use in the promotion of natural knowledge, yet the discovery of simple and uniform principles, by which a great number of apparently heterogeneous phenomena are reduced to universal laws, must even be allowed to be of considerable importance towards the improvement of the human intellect.[34]

After the three propositions supported by Newton, Young included seven propositions on the propagation of undulations through media, the uniform velocity of undulations and the laws of the reflection and refraction of undulations, supported mainly by Huygens, Euler, LaGrange and his own 1800 paper. His arguments were generally a restatement of the views he had expressed on these topics in his 1800 paper, but now concerned specifically with light. Again he took great care in answering Newton's objections to the possibility of rectilinear light propagation without great divergence. He provided a reinterpretation of the argument, used by Newton, that undulations would diverge in fluids, just as sound diverged in air. Rather than concluding that light would diverge, Young wrote:

As to the analogy with other fluids, the most natural inference from it is this: "The waves of the air, wherein sounds consist, bend manifestly, though not so much as the waves of water; water being an inelastic and air a moderately elastic medium; but ether being most highly elastic, its waves bend very far less than those of the air, and therefore almost imperceptibly."[35]

The law of interference applied to light was set forth in proposition eight.[36] He followed this proposition of interference with two important corollaries: "(1) Of the colours of striated surfaces and (2) of the colours of thin plates." In these two corollaries, Young saw the best confirmation of the whole undulatory theory. He wrote of his explanation of striated surfaces:

This experiment affords a very strong confirmation of the theory. It is impossible to deduce any explanation of it from any hypothesis hitherto advanced; and I believe it would be difficult to invent any other that would account for it.[37]

With this assertion, of course, he rejected Henry Brougham's two optical papers.

To conclude his paper, Young finally asserted that "Radiant light consists in Undulations of the luminiferous Ether." He again cleverly used Newton in an attempt to secure the acceptance of his idea by overstating Newton's support:

It is clearly granted by Newton [he wrote] that there are undulations, yet he denies that they constitute light; but it is shown in the first three Corollaries of the last Proposition [1. Of the Colours of striated surfaces, 2. Of the Colours of Thin Plates, 3. Of the Colours of Thick Plates] that all cases of the increase or diminution of light are referable to an increase or diminution of such undulations, and that all the affections to which the undulations would be liable, are distinctly visible in the phenomena of light; it may therefore be very logically inferred, that the undulations are light.[38]

Young thus presented his theory in such a way that it was impossible to deny either that it was extremely useful in the explanation of several formerly inexplicable phenomena, or that Newton himself had at times appeared to hold views similar to those of the undulatory theory. His explanation of thin films, thick films, and striated surfaces, based on the law of interference, was unquestionably of value and his explanations of reflection and refraction were mathematically supported by Euler, Huygens, and LaGrange.

By 1802, Young had succeeded in presenting an alternative theory for the explanation of optical phenomena to replace the accepted Newtonian system. The arguments Young brought forward to support his new undulatory theory were based largely on his own experiments on sound performed at Cambridge and his use of quotations from the revered works of Newton. He lacked actual optical experiments, however, to completely support his new theory. Young provided this experimental support, for both the undulatory theory in general and the law of interference in specific, in a new paper entitled "Experiments and Calculations relative to physical Optics." Young's presentation took the form of the Bakerian Lecture given before the Royal Society on November 24, 1803. For the first time, the results of "traditional" optical experiments were given in support of the applicability of his law of interference. These were experiments using the standard equipment of sunlight, holes in screens, prisms, hairs, and darkened rooms. His purpose for the various experiments was to demonstrate that the seemingly different phenomena of inflection and thin plates could all be included under one description, using the new law of interference. Young described various experiments, including those in which he used hairs set in the path of a beam of sunlight to produce inflection (diffraction) patterns. After careful observations and measurements of these patterns, giving tabulated results, he concluded that the pattern, or rather the spacing of the lines of the pattern, produced by the hair's diffraction of light was closely proportional to the pattern produced by the air between two plates. Young concluded therefore:

It is very easily shown, with respect to the colours of thin plates, that each kind of light disappears and reappears, where the difference of the routes of the two of its portions are in arithmetical progression; and we have shown that the same law may be in general inferred from the phenomena of diffracted light, even independently of analogy.

The distribution of the colours is also so similar in both cases, as to point immediately to a similarity in the causes.[39]

Young was, however, primarily interested in demonstrating the descriptive value of the undulatory theory, without introducing any statements involving causes. In fact, by this time, Young had realized the impossibility of establishing any causal explanation for the undulatory theory. He changed his direction in this paper by establishing a "descriptional" basis for the system. He retracted his tentative proposals about ether density as a cause, put forth as a temporary scheme in his 1800 and 1802 papers. "I have not," he wrote,

in the course of these investigations, found any reason to suppose the presence of such an inflecting medium in the neighborhood of dense substances as I was formerly inclined to attribute to them. . . .[40]

Young was led to this renunciation of ether density partly, it seems, from his inability to formulate a definite description of how it worked, and partly from consideration of Bradley's experiment on the aberrations of starlight. He wrote:

And, upon considering the phenomena of the aberration of the stars, I am disposed to believe, that the luminiferous ether pervades the substance of all material bodies with little or no resistance, as freely perhaps as the wind passes through a grove of trees.[41]

The passage of the earth through the ether with little or no hindrance could not be reconciled with any explanation of inflection in terms of increased ether density close to material bodies. If material bodies did collect and somehow maintain a greater density of ether close to their surfaces, it was hard to see how there could be free passage of the earth through the ether without interaction with the ether. The aberration of starlight was an unexplained problem for the undulatory theory. To increase the embarrassment, it had a very simple explanation in terms of the corpuscular theory.[42]

Young was clearly most interested in presenting his new idea of interference, with appropriate experimental support. Problems of explanation in terms of an ether were of secondary importance. A second aspect of this paper was Young's direct criticism of the Newtonians and the emission theory. He used the experiments and measurements on diffraction patterns as the basis for his criticism, stressing the success of the undulatory theory in describing these phenomena. Young referred to the experimental work of Grimaldi to illustrate his remarks on diffraction. Grimaldi's work contained many experiments on the effects of diffraction. He had been particularly interested in the spreading of light as it entered a dark chamber through small holes, and in the colored fringes which generally accompanied the beam of light, formed in the shadows around the circle of light. In certain cases, Grimaldi had observed that two circles of light produced by small holes, when superimposed, appeared to partially destroy each other, within the overlapping parts of the circles. Grimaldi had not pursued an explanation of the various diffraction effects. He had been more concerned with determining whether light "was a material or an accident."[43]

Young asserted that his new law of interference could be used to offer a very simple explanation for all of Grimaldi's observations. He challenged the Newtonian system to do as well:

The experiment of Grimaldi on the crested fringes within the shadow, together, with several others of his observations, equally important, has been left unnoticed by Newton. Those who are attracted to the Newtonian theory of light, or to the hypotheses of modern opticians, founded on views still less enlarged, would do well to endeavour to imagine anything like an explanation of these experiments, derived from their own doctrines and if they fail in the attempt, to refrain at least from idle declaration against a system which is founded on the accuracy of its application to all these facts, and to a thousand others of a similar nature.[44]

This was a direct challenge to the old system of optics to compete with a new interpretation offering a different perspective. Young defied his opponents either to produce a more complete explanation of these long neglected phenomena based upon their own theory, or to refrain from their non-constructive criticism of the new concept which had demonstrated applicability. His challenge to Newtonian optics struck at its weakest place.

Young built his case skillfully. He managed to include the newly discovered phenomenon of "dark rays" within the range of explanation offered by his new law of interference. In 1801, J.W. Ritter[45] discovered that rays beyond the violet end of the spectrum produce a chemical effect on silver nitrate. The "dark rays," as they were called, seemed to extend beyond the violet end of the spectrum for a distance equal to that which was occupied by the violet portion of the spectrum itself. Young was interested in seeing whether his undulatory theory could be used to describe these rays, and was especially eager to discover whether the law of interference applied to the dark rays as well as to visible light. He conducted an experiment on the rays, trying them in a thin plate of air to see if they produced interference rings in a way similar to visible light.

For this purpose [he wrote] I formed an image of the rings (from sunlight), and I threw this image on paper dipped in a solution of nitrate of silver.... In the course of an hour, portions of three dark rings were very distinctly visible, much smaller than the brightest rings of the colored image, and coinciding very nearly in their own dimensions, with the rings of violet light that appeared upon the interposition of violet glass. I thought the dark rings were a little smaller than the violet rings, but the difference was not sufficiently great to be accurately ascertained.... [46]

Young found that these "dark rays" seemed to display the same properties as rays from the visible spectrum, and that, most significantly, they could be easily described by using the law of interference. Based on his success with "dark rays," Young proposed that a similar experiment could be performed on the rays of invisible heat discovered by William Herschel in 1800. He was sure that these rays would also lend themselves to an easy description by the law of interference.

If we had thermometers sufficiently delicate [Young wrote], it is probable that we might acquire, by similar means, information still more interesting with respect to the rays of invisible heat discovered by Dr. Herschel, but at present, there is great reason to doubt the practicability of such an experiment.[47]

Being confident that the "invisible rays of heat" would obey the law of interference, Young was sure they could be explained in terms of the undulatory system. This possibility seemed especially important to Young because he hoped that the undulatory theory could then be related to

Count Rumford's new theory of heat, using the phenomena of "invisible rays" as a starting point. Although Count Rumford said little concerning the undulatory theory, Young was confident that Rumford's theory could somehow be directly related to the undulatory theory. If this relationship could be shown, it would have been a striking theoretical extension of the undulatory theory. Young simply suggested this as a possibility for the future.

There were several important supports for Young's undulatory theory in the early years. The theory's demonstrated compatibility with the newly discovered phenomena of ultra-violet and infra-red certainly helped. There were, in addition, two significant experimental supports for the theory, supplied in 1802 by Sir Henry Englefield and William Hyde Wollaston.

Sir Henry Englefield was fascinated by Herschel's articles[48] on the newly discovered phenomena of "invisible heat." Pursuing suggestions by Herschel, he performed a number of experiments on the separability of "invisible heat" and light by refraction. Sir Henry took this separability of heat and light to be a confirmation of the undulatory theory of light. He concluded that the results of his investigations "carried with them . . . a complete conviction of the truth and accuracy of the Doctor's [Young's] assertions. . . ."[49]

Another investigation by Sir Henry provided a more indirect support for Young's theory. In July 1802, he published an experiment "On the Effect of Sound upon the Barometer."[50] He noticed that every time a large church bell was struck, a barometer located near it would show a rise in the barometric column. Sir Henry could offer no explanation for this observation. Young, however, was able to offer a nice explanation by means of the coalescence of sound waves.[51]

William Hyde Wollaston was also favorably disposed toward Young's new undulatory theory. Wollaston was interested in the refractive and dispersive properties of various substances. To investigate these properties, Wollaston employed a new method of investigation which enable him to study substances whose properties had not previously lent themselves to study. This new method, Wollaston wrote:

Is not only convenient in common cases of refraction, but also capable of affording results not obtainable by other means.

"This method was suggested by a consideration of Sir Isaac Newton's prismatic eye glass, the principle of which depends on the reflection of light at the inner surface of a dense refracting medium."[52]

Wollaston reasoned that since the range within which the total reflection takes place depends upon both the density of the reflecting prism and upon the rarity of the adjacent medium, the extent of the range should vary with the difference in densities of the two media. Therefore, if the refractive power of one medium is known, the refractive power of any other less dense medium can be determined by measuring the angle at which it reflects a ray of light. Wollaston determined the refractive properties of a great many substances, among them Iceland crystal.

The importance of Wollaston's work as a support for Young's new theory became clear in his second paper "On the oblique Refraction of Iceland Crystal."[53] Wollaston made a detailed study of the properties of Iceland crystal. At the beginning of his paper, he wrote:

The optical properties of this body have been so simply described by Huygens, in his *Traite de la Lumiere*, that it could answer little purpose to attempt to make any addition to those which he has enumerated. But, as the law to which he has reduced the obliqued refractions occasioned by it, could not be verified by former methods of measurement, without considerable difficulty, it may be worth while to offer a new and easy proof of the justness of his conclusions. For since the theory by which he was guided in his inquiries, affords (as has lately been shown by Dr. Young*) a single explanation of several phenomena not yet accounted for by any other hypothesis, it must be admitted that it is entitled to a higher degree of consideration than it has in general received."

*Bakerian Lecture. *Phil. Trans.* for 1801.[54]

Wollaston found that the undulatory theory predicted results for the double refraction phenomena which were in very close agreement with the values which he had determined by experiment. Since he had begun his investigations of Iceland crystal before Young published his Bakerian Lecture, he was, therefore, impressed by Young's attempted re-assertion of an undulatory theory, and felt that this theory should at least be given a hearing. Furthermore, Wollaston concluded that the separation of the two refracted rays by Iceland crystal, predicted by the undulatory theory, was in fact the separation of the rays measured by his experiments.

Young's assertion of the undulatory theory appeared to be achieving acceptance in 1802. The theory had demonstrated its usefulness in explaining a large number of optical phenomena. The new law of interference made it possible to group the previously diverse phenomena of inflection and the colors of thin plates together under one simple description. In addition, the theory had received some support from two respected experimenters. But this attempted establishment of the undulatory system stood in direct opposition to the respected Newtonian system of optics. The undulatory theory clearly could not be reconciled with a system of optics based on forces and light corpuscles. More specifically, the new system conflicted with the optical writings of contemporary Newtonians. Just as there was opposition raised against Young's theory of sound, criticisms of Young's theory of light quickly appeared. Foremost and most vehement of the critics was the young Henry Brougham. Young's law of interference applied to light was a direct contradiction to Brougham's first two scientific papers. Young swiftly realized that the Newtonian system of optics would prove difficult to replace.

Henry Brougham's Attack

Henry Brougham saw himself as the self-styled defender of the Newtonian tradition in all its purity. In 1802, a group of men including Sydney Smith, Francis Jeffrey, and Henry Brougham started what was to be a notably influential magazine, *The Edinburgh Review*. This *Review* became one of the most influential of several magazines which sprang up in the early nineteenth century, opening up a whole new critical dimension to the learned world. A contemporary wrote of the *Review*:

> The effect was electrical. It was an entire and instant change of everything that the public had been accustomed to in that sort of composition. The learning of the new journal, its talent, its spirit, its writing, its independence, were all new. Its literature, its political economy and its pure science were generally admired. It was hailed as the dawn of a bright day.[55]

Another commentator wrote:

> Now was beginning the race of reviews and magazines, feeble precursors indeed of the periodical hosts of our days, but already powerful for good and evil, able to wound, but on the whole serving to keep together the commonwealth of letters, and to make its influence felt.[56]

Brougham launched his attack on Young and the undulatory system in the first issue of the *Edinburgh Review* and reached a peak of malice difficult to describe. His criticisms were almost entirely "ad hominum," not seeking to refute Young's work, but to discredit both Young and his theory in any, and every, way possible. Brougham was probably astute enough to realize that if the new theory of light became established, it would constitute a categorical refutation of his own work and of the corpuscular theory of light in general. It would also be a serious breach in the established Newtonian structure of science as Brougham envisioned it. For him, science consisted of what he considered to be the Baconian method of pure induction on the one hand, and the dogmatic application of the concepts contained within the *Principia* on the other.[57]

Several quotations from Brougham's reviews will serve to show the extent to which he was willing to go in his attack, as well as to reveal his attitude toward traditional English science and the Royal Society. Of Young's Bakerian Lecture of 1801, Brougham wrote:

As this paper contains nothing which deserves the name, either of experiment or discovery, and as it is in fact destitute of every species of merit, we should have allowed it to pass among the multitude of those articles which must always find admittance into the collections of a Society which is pledged to publish two or three volumes every year. The dignities of the author and the title of Bakerian lecture, which is prefixed to these lucubrations should not have saved them from a place in the ignoble crowd. But we have of late observed in the physical world a most unaccountable predilection for vague hypotheses daily gaining ground; and we are mortified to see that the Royal Society, forgetful of those improvements in science to which it owes its origin, and neglecting the precepts of its most illustrious members, is now, by the publication of such papers, giving the countenance of its high authority to dangerous relaxations in the principles of physical logic. We wish to raise our feeble voice against innovations, that can have no other effect than to check the progress of science, and renew all those wild phantoms of the imagination which Bacon and Newton put to flight from her temple.[58]

Brougham also added gratuitous assumptions to his argument:

Were we to take the trouble of refuting him, he might tell us, "My opinion is changed, and I have abandoned that hypothesis; but here is another for you." We demand if the world of science which Newton once illuminated is to be as changeable in its modes, as the world of taste, which is directed by the nod of a silly woman, or pampered fop? Has the Royal Society degraded its publications into bulletins of news and fashionable theories for the ladies who attend the Royal Institutions? Proh pudor! Let the Professor continue to amuse his audience with an endless variety of such harmless trifles; but in the name of science, let them not find admittance into that venerable repository which contains the works of Newton, and Boyle, and Cavendish and Maskelyne and Herschel.[59]

His continued criticism of Young's later papers was in the same vein:

The paper which stands first [Bakerian Lecture 1804] is another Bakerian Lecture, containing more fancies, more blunders, more unfound hypotheses, more gratuitous fictions, all upon the same field on which Newton trode, and all from the fertile, yet fruitless brain of the same eternal Dr. Young.[60]

Not only was Young attacked, but the work of Wollaston and Rumford also came under Brougham's fire. In April, when speaking of Young and Wollaston, he wrote rather ironically:

> ... These authors, misled by the nature of sound, do not admit the materiality of light, but assert that it is a vibration propagated through the medium. But, short as these remarks are, we are loath to waste any more time on such a feeble and ill-conducted defense of an untenable and useless hypotheses.[61]

Wollaston's experimental work seemed to lend support to the simplicity and usefulness of Young's law of interference. Brougham did not hesitate to attempt to discredit Wollaston's work.

Even Count Rumford's new theory of heat was not spared criticism in the *Review*. Brougham recalled Young's suggestion of the possible relation between the two new theories and was careful to criticize them both.

> ... A paper filled with theoretical matters [he wrote of Thomson's paper], abounding in pulses, vibrations, internal motions, and etherial fluids, deserves to be exposed; the more, because these chimeras are mingled with a portion of induction, and have received the ill-deserved honour of a place in the Philosophical Transactions.[62]

The bitterness and persistence of Brougham's attacks are difficult to explain. There were, however, several factors which may have contributed to Brougham's biliousness. In 1800, in an essay on cycloidal curves, Young wrote disparagingly of an article written by Brougham in the *Philosophical Transactions* for 1798.[63] Young's remarks appear to be justified, but they were stated with unfortunate bluntness. Young pointed out that the amount of material published in the form of mathematical demonstrations was tremendous indeed. This material was scattered among various journals in no particular order as to subject. Some system of ordering mathematical publications so that various mathematical demonstrations could be indexed and made easily available to a person beginning his study of mathematics was sorely needed.

Until this is done [he wrote] nothing is left but for every individual who is curious in the search of geometrical knowledge to look over all the mathematical authors and all the literary memoirs of the last past and present centuries: for without this he may very easy fancy he had made discoveries, when the same facts had been known and forgotten long before he had existed. An instance of this has lately occurred to a young gentleman in Edinburgh, a man who certainly promises in the course of time, to add considerably to our knowledge of the laws of nature.[64]

Then, referring to specific examples of previous demonstrations, Young wrote:

On the whole it appears that this ingenious gentleman has been somewhat unfortunate in the choice of these problems which he has selected as specimens of the elegance of the modern mode of demonstration; whether those which he has brought foreward without proof would have furnished him with a more favorable opportunity for the display of neatness and accuracy, may be more easily determined whenever he may think proper to lay before the public their analysis, construction, and demonstration, at full length.[65]

Young's remarks were rather deflating to the "young man from Edinburgh," who had thought he was presenting a new, clear method of analytical demonstration of some new problems in mathematics. In this instance, Young took the side of the traditional geometrical demonstrations of these problems, claiming that Brougham's work was not only repetitive of previous work, but was inferior to the older demonstrations. He also implied by the phrase "in the course of time," that Brougham's previous two papers on optics had done nothing to add to "our knowledge of the laws of nature."

Second, Young's papers in the *Philosophical Transactions*, asserting the undulatory theory and the law of interference, stood in direct contradiction to Brougham's papers on optics, also published in the *Philosophical Transactions*. Brougham was an avowed adherent of the emission theory of light. His explanations of the phenomena of optics were therefore expressed in terms of forces and corpuscles of light. Brougham clearly thought that he had significantly extended the theory of optics by his papers. In particular, his explanation of the colors of striated surfaces he considered to be of special importance. It was posed to offer an explanation of this phenomenon in terms of forces and corpuscles. Young's work, therefore, not only conflicted with the traditional Newtonian explanation of optics, but it explicitly denied Brougham's new work on striated surfaces.

Young himself, as we have seen, considered his own explanation of the colors of striated surfaces a great support for his law of interference. But of the phenomenon itself, Young wrote:

Boyle appears to have been the first to observe the colours of scratches on polished surfaces. Newton has not noticed them. Mazeas and Mr. Brougham have made some experiments on the subject, yet without deriving any satisfactory conclusion.[66]

Young's explanation, of course, was in terms of interference (see Fig. 8). It was:

Let one of the points be now depressed below the given plane; then the whole path of the light reflected from it, will be lengthened by a line which is to the depression

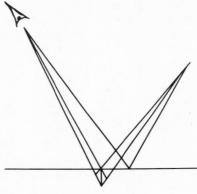

Fig. 8 Interference at striated surface.

of the point as twice the cosine of incidence to the radius. Fig. 2.

If, therefore, equal undulations of given dimensions be reflected from two points, situated near enough to appear to the eye but as one, wherever this line is equal to half the breadth of a whole undulation, the reflection from the depressed point will so interfere with the reflection from the fixed point, that the progressive motion of the one will coincide with the retrograde motion of the other, and they will both be destroyed; but when this line is equal to the whole breadth of an undulation, the effect will be doubled and when to a breadth and a half, again destroyed, and thus for a considerable number of alterations. . . .[67]

Young's new explanation of the phenomenon was much simpler than Brougham's. Furthermore, Young's explanation did not constitute an addition to his system superimposed especially to account for a new phenomenon. Brougham had to devise a complex, new process of color formation by reflection to account for the new observation of colors from striated surfaces. Young simply extended his law of interference to easily include the new phenomena.

Aside from Young's derogatory comment about Brougham's work on striations, the new undulatory theory called the whole basis of Brougham's work into serious question. Brougham's two papers were based on the assertion that the particles of light were of various sizes, distinguishable by the color they produced. These variously sized light corpuscles were assumed by Brougham, as by Newtonians generally, to have equal velocities, regardless of their size or of the source which emitted them. Young objected specifically to these assumptions, considering them unfounded and unsupportable. Instead, Young presented a system of optics which offered an easy explanation both for the equal velocity of the propagation of light and for the equal velocity of emission of light from all light sources. Brougham could not defend his assumptions either with experiment or sound reasoning. Young's objections were, therefore, severe criticisms of Newtonian optics, which, unanswered, cast serious doubt on the whole system.

Young's own criticism and works might have accounted for the bitterness of Brougham's attack if Brougham were a person disposed to quick anger and revenge. But they are less satisfactory in accounting for the attack on Rumford's theory of heat and on Wollaston, a respected experimenter. The reason behind the wide range of Brougham's attack seems to have been his veneration of Newton and what he considered to be the Newtonian tradition. We have seen examples of this veneration in Brougham's praise of Newton, in his reluctance to associate anything but a corpuscular theory of light with the *Opticks*, and in his concern for the purity of this tradition in the Royal Society. His claimed esteem of Bacon's works on method and of Newton's methods of experimentation was obvious in his two optical papers. His experiments and descriptions were models of the way in which Newton had presented his work. Brougham saw in Young's new system of optics a threat to the Newtonian tradition. Young had not performed the traditional experiments in optics to reach his conclusions. He did not use beams of sunlight, prisms, and knife blades but perversely experimented with sound and air flowing through tubes, organ pipes, and smoke streams. Brougham considered this a most untraditional approach to "Newton's subject" of optics.

But if the approach was untraditional, the basis of the new system was even worse. It was disconcerting to have Newton's speculations on "Fits" and pulses in the ether brought to the fore after almost a century of neglect and then related to the new undulatory theory. Young used all of the sections and phrases from Newton's *Opticks* that the Newtonians had attempted to play down for a century. The establishment of waves and an ether in a realm formerly explained in terms of corpuscles and forces would be a major embarrassment to the Newtonian tradition of optics.

It seems, therefore, that Wollaston was attacked by Brougham because his experiments lent support to the law of interference. Rumford's theory of heat was criticized because Young himself hoped to find support for his system in this new theory of heat. The scope of Brougham's attack becomes explicable in light of his concern for the defense of the Newtonian tradition as a whole.[68] We will see this same attitude recurring in the works of David Brewster.

There can be no doubt of the effectiveness of Brougham's attack on

Young, whatever his motives. Young wrote a reply to his criticism, but it was long, detailed, and sold only one copy.[69] Brougham had clearly carried the day. George Peacock wrote of the *Edinburgh Review* articles:

Their influence, however, upon public opinion was more remarkable than could reasonably have been expected, even from the great authority of the publication in which they appeared, and the unquestionable ability with which they were written. They not only seriously damaged, for the time, the estimation of the scientific character of Dr. Young, but diverted public attention from the examination of the truth of his theories, at least among his own countrymen, for nearly twenty years.[70]

George Ellis, a contemporary of Young, wrote the following to Sir Walter Scott of Young's attempt to publish his lectures to the Royal Institution:

He was actually preparing for press, when the bookseller came to him and told him that the ridicule thrown by the Edinburgh Review on some papers of his in the Philosophical Transactions had so frightened the whole trade that he must request to be released from his bargain.[71]

The *Review* had an unquestionable effect on Young's work in science. Young did, in fact, withdraw from scientific controversy after 1804. But his withdrawal was occasioned not because of the damage to his scientific reputation, but because of the possible damage to his professional career as a doctor. It must be borne in mind that Young's primary professional purpose in life was to be a good doctor. His whole education had been undertaken for the purpose of becoming an accomplished physician. In his reply to criticism, Young wrote:

... For the talent which God has not given me, I am not responsible, but those which I possess, I have hitherto cultivated and employed as diligently as my opportunities have allowed me to do; and I shall continue to apply them with assiduity, and in tranquility, to that profession which has constantly been the ultimate object of all my labors.[72]

Both Young and his friends thought it advisable for him to withdraw from public scientific activity because of the bad effect on his medical career. Young's friend, Hudson Gurney, wrote of this period:

At this time he resigned his office as lecturer to the Royal Institution, it being thought by his friends that his holding it longer would be likely to interfere with his success as a medical practitioner. This view, as regarding his continuance in a situation which would appear to the public as anomalous to his profession and hardly compatible with its duties, was probably a just one.[73]

But this view, taken by all Young's biographers, that the withdrawal came about because of a probable hindrance to his work as a physician is one-sided and a bit too favorable to Young. It was true that Young withdrew from public scientific life. However, his withdrawal was not as drastic as it would appear. Young, in fact, enjoyed only limited success as a lecturer in the Royal Institution. His lectures covered the whole realm of physical science, at a depth considerably greater than the majority of the members of his audience could cope with, his style was laborious and he did not mold the lectures into interesting presentations. At the time of Young's resignation, he had completed two lecture series. The first series consisted of thirty lectures, the second was expanded to sixty-two lectures. The substance of this last series resulted in Young's published *Lectures on Natural Philosophy*.[74] There is ample evidence to indicate that Young had advanced himself and his lectures to a point where resignation did not represent the cancellation of any work or interest. His time of public resignation came at a natural stopping place.

There is also evidence to indicate that Young had presented a rather complete account of his achievements in optics in his papers to the Royal Society. Here again, it does not appear that he was forced to abandon any work in progress. Young seemed to have completed his researches. He wrote to Dalzel in March 1802:

The theory of light and colours, though it did not occupy a large portion of time, I conceive to be of more importance than all that I have done, or ever shall do besides.[75]

This letter lends weight to the conclusion that Young had published his theory of light and colours in a complete form by 1804. It did not, in fact, take a great deal of his time, not nearly as much as his preparation for the *Lectures in Natural Philosophy*. Young certainly continued his interests in optics in his leisure time, but in private and by presenting them for anonymous publication to journals like the *Quarterly Review* and *Nicholson Journal*. He also felt free to continue open publication of articles on topics directly related to medicine.

The main effect of Brougham's criticism lay, therefore, in the effect on public acceptance of the undulatory theory in England. Men no longer referred to Young's theory after 1804, as Englefield and Wollaston had done before. Public support for the undulatory theory disappeared after the criticisms in the *Edinburgh Review*. The undulatory theory had seemed well on its way to replace the Newtonian emission theory before 1804, but the conservative forces were massed against its acceptance after the criticism. The years 1804–1820 saw the increase in popularity of the emission theory in England, rather than its gradual replacement by the newer concept. This increased popularity was caused by a number of factors. The undulatory theory was by no means overwhelmingly better than the Newtonian system. True, it was simpler and clearer in some respects, but there were no decisive experiments nor arguments which would decide the question. Furthermore, the emission theory possessed the secure position of tradition and compatibility with the accepted concepts of the physical world. Newton and his works, however modified by his Newtonian followers, were held in great esteem by the English. Second, Brougham's severe criticism forced Young to withdraw from active public support of the undulatory theory after 1804. The theory was left without a public champion and suffered accordingly.[76] Third, and perhaps most important, the continued success of Newtonian optics in England during this period was largely the result of the new phenomena of polarization and the work of David Brewster.

IV
The Tenacity of Newtonian Optics in England

David Brewster, the Last Champion

David Brewster became the champion of the Newtonian system of optics in England during the first three decades of the nineteenth century. Brewster was firmly committed to the explanation of all optical phenomena in terms of forces and light particles. His interest in optics dated from his early education, when he began to build telescopes and other optical instruments. Brewster was familiar with Thomas Young's work in optics and he recognized that the new concept of interference was of possible use for the explanation of various optical phenomena. But this recognition was by no means an admission of the inferiority of the corpuscular interpretation of those same phenomena. Brewster's early commitment to the Newtonian system was reinforced in 1809 by the works of Laplace and Malus in France, and by Thomas Young's inability to promptly and convincingly counter these works.

Pierre Simon Laplace published a paper in January, 1809[1] which attempted to explain the difficult problem of double refraction in terms of light corpuscles and forces of attraction and repulsion. Laplace treated the difference between the ordinary and the extraordinary rays produced by the crystal in terms of the differing velocities of the corpuscles composing these rays. The crystal was supposed to exert different forces upon the light corpuscles which composed the ordinary and extraordinary rays, in a ratio which depended upon the inclination of the extraordinary ray to the axis of the crystal. The difference of the squares of the velocities of the ordinary and extraordinary rays were calculated to be proportional to the square of the sine of the angle which the extraordinary ray made with the crystalline axis. Laplace then applied Fermat's principle of least time to the phenomenon and discovered that the results of the calculations on velocities of the rays agreed well with the observations on doubly refracting crystals. This was a detailed attempt to extend the use of corpuscles and forces to include double refraction.

Thomas Young was especially interested in Laplace's paper, particularly because Laplace had ignored his own explanation of double refraction in terms of the undulatory theory. Laplace also chose to ignore Wollaston's experimental support of Young's theory on double refraction. Laplace's paper represented a standard technique with regard to Thomas Young's papers. He simply avoided mention of them and proceeded as if the undulatory theory could not supply an explanation as suitable as one involving corpuscles and forces.[2]

Young responded anonymously to Laplace's paper in a review for the *Quarterly Review* published in November 1809.[3] This review illustrates the handicap which Young's forced anonymity imposed on his defense of the undulatory theory in England. He could not refer to his own work openly, nor could he convince anyone else to support the undulatory theory publicly. Consequently, his criticisms of Laplace's paper, while justified and important, were ineffectual.

Laplace's work was greatly reinforced in 1809 by Malus' discovery of polarization. Etienne Louis Malus was awarded the prize of the Académie des Sciences in 1810 for his memoir "Theorie de la double refraction de la lumiere dans les substances cristallisee." This memoir began with a comparison statement of the theories of Huygens and Newton. Malus declared that he would adopt the emission theory as his starting point in his investigations because the theory of vibrations in an ethereal fluid was not compatible with certain chemical phenomena produced by light. His memoir was written in response to a prize offered in 1808 on the topic: "To give a mathematical theory supported by experiment, of the double refraction which light undergoes in traveling through various crystals."

Malus's work addressed itself to two difficult problems for corpuscular optics, the phenomenon of double refraction itself and the problem of partial reflection at the surface of substances. His memoir specifically related the two phenomena of double refraction and polarization by reflection.[4]

Thomas Young was forced to admit, anonymously, in the *Quarterly Review*,[5] that Malus's discovery raised grave difficulties for the undulatory theory. He was forced to argue weakly that even though Malus's work appeared to demonstrate conclusively "the insufficiency of the undulatory theory, in its present state, for explaining all the phenomena of light," it did not therefore mean the "perfect sufficiency of the projectile system." Much more evidence was needed, Young argued, before a positive decision could be made concerning the superiority of the undulatory or the corpuscular theory of optics.

Young was certainly not prepared to accept the defeat of the undulatory theory by this one new phenomena. But polarization did, in fact, present an insurmountable problem until 1819. Malus was very clear in his suggestion that phenomena of polarization which he observed could be best explained in terms of a polarity of the particles of light; in fact, just the type of polarity which Newton had suggested. Malus presented an explanation in terms of the positions of the axes of the particles of light; he considered this explanation sufficient. Whenever the axes of the light particles were oriented in a certain way, those special particles would be reflected from the surface of a reflecting body. Those particles with their axes oriented in different directions were not reflected but rather transmitted. Even Young was forced to admit:

It seems to be undeniable, that the general tenour of these phenomena is such, as obviously to point at some property resembling polarity, which appears to be much more easily reconcilable with the Newtonian ideas than with those of Huygens.[6]

The problem of polarization baffled Young.

David Brewster immediately began an extensive study of polarization phenomena after hearing of Malus's discovery. Brewster agreed with Malus that the new phenomena of polarization could *best* be explained in terms of the polarity of light particles. He began at once to investigate the phenomena in detail, and in so doing provided further elaboration for the similarity of the processes of double refraction and polarization. Brewster realized that the two rays produced by double refraction behaved in a manner similar to the polarized rays produced by reflection.

The undulatory theory was, therefore, squarely confronted by two unanswerable problems instead of just one. The explanation of double refraction, using longitudinal impulses, had always been one of the strong points of the undulatory theory. Now, the old explanation of double refraction offered by the undulatory theory could no longer be easily maintained because it could not be related in any way to polarization by reflection. The undulatory explanation of double refraction involved the formulation of an elliptical and a circular shape to the wave front by the crystal, which produced the ordinary and extraordinary rays. The characteristics of the rays of double refraction were caused by the different indices of refraction in the crystal for the different elliptically and circularly shaped pulses. There was no way that this explanation of double refraction could be related to the polarized ray produced by reflection. The undulatory theorists in England, specifically Young, were unable to advance another timely explanation for the polarity of the extraordinary ray of double refraction or for the polarity of the ray produced by reflection. By contrast, the emission theorists were quick to offer an explanation. Newtonian optics owed its continued life during the nineteenth century in large part to its ability to offer a quick, plausible explanation of the problem of polarization.

David Brewster was the man most responsible for the success which the Newtonian system enjoyed in its explanation of polarization. He published his first paper on this subject in 1813. Brewster began his researches by repeating Malus's early experiments, with good agreement. But when he took a thin sheet of agate and viewed objects through it, he observed that there was a colored fringe produced on both sides of the objects.

Upon examining this colored image with a prism of Iceland spar [Brewster wrote], I was astonished to find that it had acquired the same property as if it had been transmitted through a double refracting crystal, and upon turning the Iceland spar about its axis, the images alternately vanished at every quarter of a revolution.[7]

Concluding that the "light polarised by transmission comported itself, in every respect, like one of the pencils formed by double refraction,"[8] he noticed that the obliquity of the angle of refraction had an effect on the polarization. "The discovery, however, of the polarization of light by oblique refraction, forms the connecting link between these two classes of facts, and holds out the prospect to obtaining a direct explanation of the leading phenomena of double refraction."[9] This discovery of polarization by transmission proved the stimulus for Brewster's subsequent researches. He continued to investigate the properties of agate and, in doing so, began an interest in the relationship between crystal structure and optical activity.

It was in the study of the effects of transmission that Brewster first hinted at an explanation of the phenomenon. His work in optics before this had taken the form of strictest adherence to pure mathematical and experimental description without any attempt at explanation. Brewster's papers and, in fact, a whole book entitled *Treatise on New Philosophical Instruments* published in 1813, had been purely descriptive, with no attempt at explanation of the observed phenomena. Now, a hope for physical explanation was explicitly stated. There were two sources for Brewster's expectations at this point in his development. One was the Newtonian tradition of forces acting at a distance on particles of light, and the other was his new interest in crystal structure. These two interests combined in Brewster's work, resulting in the culmination of the development of the Newtonian system of optics.

Brewster's adherence to the Newtonian tradition of physical optics became clear in his 1814 paper entitled "On the Properties of light exhibited in the Optical Phenomena of Mother of Pearl. . . ." In this paper, he tried to find the cause of the optical properties of the striated surface. The colors of striated surfaces had been an interesting problem in optics. There were two possible explanations for these colors, offered by the Newtonian system and by Thomas Young. Brewster decided to ignore Young's application of the law of interference to the colors of striated surfaces and to refer back to the two papers on this subject presented by his old schoolmate, Henry Brougham. Brougham had applied his modification of the zones of forces, and postulated the formation of the colors on the mother-of-pearl shell by refraction, as a result of these forces on light corpuscles. Brewster expanded upon this Newtonian interpretation to produce what he considered to be the complete explanation of the color phenomenon. A long quotation from Brewster will serve both to describe his explanation and to give an example of this use of forces and corpuscles in the Newtonian tradition of optics (see Fig. 9):

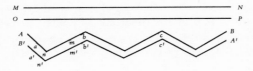

Fig. 9   Forces above surface of mother of pearl.

. . .Let us now suppose that fig. 5 represents a section of a plate of mother of pearl, having *AanmbcB* for its upper surface, and *B'a'n'm'b'c'A'* for its lower surface. Let *OP* be the line at which the attracting or reflecting force begins, and let the reflecting force terminate at *MN* according to the Newtonian theory. We have already seen that when the surface *Ab* of the mother of pearl is ground as flat as possible, and brought to a high polish, the light which is incident on the repulsive stratum *MNOP* is reflected as in all other bodies, and affords a perfect image of the object from which it radiates. Hence, it follows that the light which forms the extraordinary images has escaped reflection, and penetrated the attractive stratus *OP*, and that its separation into colours and extraordinary reflection are produced by one or more causes residing between *OP* and the surface *AB* of the mother of pearl.

Let us first attend to the aberration of the extraordinary images. Since the real surface of *AB* is composed of faces inclined to the general surface *AbcB*, we are led to suppose that the primary extraordinary image is reflected from the face *m*, while the secondary extraordinary image is reflected from the face *n*. Now this could only happen from two causes, either in consequence of the mother of pearl having a repulsive force different from the ordinary repulsive force which produces reflection; or from its possessing the power of reflecting light from its actual surface. That this extraordinary force, is, in other respects like the ordinary reflecting force, is manifest from a portion of the extraordinary pencil being transmitted, while the other portion suffers reflection. The existence of such a force being unquestionable, we have next to consider the form and position of the surfaces to which it belongs.[10]

We are now in a position to see some of the reasons behind Brewster's adherence to the emission theory. The first is found in the phenomenon of mother-of-pearl itself. The question arises, how can a striated surface produce both a normal reflection of an image, as if it were a plane surface, and the colors of the so-called extraordinary image? Young's law of interference explained the colors very well, but Brewster considered that it was unsatisfactory in describing the simple reflection because of Newton's old argument. Layers of forces seemed to him to offer a better explanation.

This seemingly tangential inquiry into pearl shells was not as divergent as it appeared. Brewster hoped to apply the Newtonian scheme of forces and particles to the phenomenon of double refraction, using an explanation of striated surfaces as a first step. This purpose was clearly revealed in one of Brewster's conclusions, posed as a query at the end of the paper:

Since a particular structure of surface is always accompanied with a new repulsive force, residing nearer the body than the common repulsive force which produces ordinary reflection, may there not reside also, near the surface of all crystallised bodies, a new refractive force which produced double refraction? And is not this supposition countenanced by the fact that the extraordinary pencil formed by Iceland spar suffers the ordinary as well as the extraordinary refraction?[11]

Brewster's line of attack was clear, although his great profusion of experiments and descriptions tended to obscure it. He wanted to apply the Newtonian theory to the phenomena of double refraction, using a new force. Brewster's work on the polarization of light by transmission and the use of Newtonian force zones led him in the direction of the applicability of forces to double refraction. But the means of applying forces to particles to produce double refraction was not evident, nor was the nature of the light particles themselves clear. Brougham had previously established their relative size, but the phenomenon of polarization demanded another property, as yet unassigned. Brewster suggested that the corpuscles of light had a new property, based on Newton's speculation that light particles had sides. But he advanced much further than mere suggestion. He put the new property of positive and negative polarization into explicit mathematical terms:

> When a beam of light is emitted by the sun, or by any other body which does not shine by reflected light, the particles which compose it are in every state of positive and negative polarization from particles completely polarized to particles not polarized at all . . . the terms positive and negative polarization being employed to denote two kinds of polarization by reflection and refraction at the polarising angle, or by reflexion in two opposite planes. A ray of direct light, before it is incident upon glass, may therefore be represented as consisting of a number of particles p, p, etc. of the following character
>
> *$p/0°$, $p/1°$, $p/2°$, $p/3°$, . . . $p/56°45'$ $-p/0°$, $p/1°$, $p/2°$, $p/3°$, . . . $p/56°45'$
>
> The particle $p/0°$ will represent a particle so completely polarized in a positive manner that it will be polarised by reflexion at $0°$ of incidence; $-p/0°$ will represent a particle so completely polarized in a negative manner that it will be polarised by reflexion in an opposite plane at $0°$ incidence; . . . $p/56°45'$ so completely unpolarized that it requires to be reflected at the maximum polarising angle before it can suffer complete polarisation.[12]

Brewster devised a system representing a beam of light particles in differing states of positive and negative polarization. These particles were acted upon by forces which lay on the surfaces of bodies. He scrupulously avoided a direct statement of "innate properties" and he was very careful not to mention or allude to Newton's "Fits of Easy Reflection and Transmission." In this respect, he was content with the whole Newtonian tradition which had attempted to neglect the "Fits."

Brewster's interest in crystal structures now assumed great importance in his work. He realized in 1816 that the idea of forces on the surface of bodies might explain polarization by reflection, but they could not explain double refraction which was a transmission phenomenon. He now attempted to relate the study of polarization to the study of the optical properties of crystals in order to explain Iceland spar. He consciously tried to relate these two studies for the next two years. His plan of attack was not direct, however, but rather, as always, very highly empirical, consisting of a great many experiments on various subjects along the way.[13]

At first glance, Brewster's work appeared to consist solely of accurate descriptions of experiments and mathematical formulations of the results. It seemed that Brewster held no commitment to either the undulatory theory or the Newtonian theories. He offered no explanations for the results of his experiments. In fact, in his own words:

In these enquiries I have made use of no hypothetical assumptions. . . . When discovery shall have accumulated a greater number of facts, and connected them together by general laws, we may then safely begin to impose better names, and to speculate respecting the cause of those wonderful phenomena which light exhibits under all its various modifications. . . .[14]

He did not, of course, do this in practice, as we have seen and as will become more and more evident. But the sheer volume of Brewster's experimental inquiry and the success of his formulation of mathematical equations to describe the phenomena, tended to obscure his underlying belief in the emission theory and his goal of bringing double refraction into the realm of Newtonian explanation.

The groundwork for the solution of the problem was found in Brewster's work on crystalline and non-crystalline bodies. He performed experiments on almost every aspect imaginable from the effects of heat on the optical properties of substances to simple index of refraction determinations. He also discovered that most substances could be made to produce double refraction under the right conditions. In short, Brewster summed up his own position by stating that "the mechanical condition of both classes of doubly refracting crystals, and the method of communicating to uncrystallized bodies the optical properties of either class, have been distinctly ascertained, and only the phenomenon which remains unaccounted for, is the division of the incident light into two oppositely polarized pencils."[15]

Brewster was baffled by this problem from February 1816 until June 1817. He was unable to find a satisfying solution using zones of force at the surface of the crystal. Although he could not reach a solution in terms of forces, he did not, however, consider the possibility of using the undulatory theory. Brewster was convinced that this rival theory could not explain the phenomenon of polarization at all. Furthermore, to his mind its applicability had been previously discredited for the problem of double refraction by two of France's great mathematicians, Laplace and Poisson. In a letter to Young in 1817, Brewster expressed his distrust of the undulatory theory:

When I mentioned to Mr. Biot, about a year ago, your demonstration, that an undulation propagated through a minutely stratified substance in which the density is greater in one direction than in another, was spheroidical, he replied that both Laplace and Poisson were of opinion that, in the present state of mathematical analysis, the simplest case of undulation could not be calculated; and therefore that the above theorem was not capable of demonstration.[16]

Young and Brewster corresponded frequently in these years between

1815 and 1817. Young attempted to convince Brewster that his law of interference was valid and should be applied, even though polarization still presented a serious problem. Neither Brewster, nor their mutual acquaintance Biot, took Young's undulatory theory seriously. Both men preferred explanations of optics in terms of corpuscles and forces. While Young had been anonymously writing review articles and wondering about polarization, men on the Continent and Brewster in England had been busily pursuing experiments on polarization from the viewpoint of Newtonian optics.[17]

With the undulatory theory completely discredited in his mind, Brewster continued to work in the Newtonian tradition. He found his solution to the problem in the reconsideration of his work on crystals and the study of the various optical axes. In 1818 he wrote:

I trust I shall be able to demonstrate, not only that the phenomena of double refraction and polarization may be explained by forces or combinations of forces different from those which have been given by Laplace and Biot, but that there are certain analogies of nature, and certain physical circumstances in the phenomena, which may lead us to select one combination of forces in preference to others, as the means which nature has employed in the accomplishment for her purposes.[18]

He concluded that forces emanating from the axes of a crystal were responsible for the action upon the light particles within the crystals themselves. Furthermore, the effect was due to one axis or to various arrangements of axes, producing equivalent effects through different configurations. "The deviation of the extraordinary ray in beryl, may be the result of a repulsive force emanating from two equal rectangular axes lying in a place perpendicular to the axis of the prism, or of various other combinations of forces, either of the same or of opposite names."[19]

Brewster drew an analogy between his work on crystal structure and the optical properties exhibited by the crystals. The action of the crystal on light was determined by the forces produced by the axes of the crystal, just as crystalline properties were determined by the crystalline axes within the crystal. The force from each axis was supposed to vary as $1/r^4$ and to have either a repulsive or an attractive effect, depending on the arrangement of the axes. To explain how certain mechanical effects could change the optical properties of a crystal, Brewster studied a great number of crystals and concluded:

The discovery of the remains of polarizing axes in a great number of crystals of this class (cubical and octohedral crystals), completely proved that, like all other crystallized bodies, they actually possess the doubly refracting and polarizing structure, but that this structure entirely vanishes in some specimens by the equilibrium of the forces in every point of the crystal, and reappears in some specimens when that equilibrium is not complete.[20]

In general, Brewster observed that most crystals had more than one axis of extraordinary refraction, not just one as previously studied. He concluded that Huygens's law for the separation of rays in Iceland spar held for crystals of one axis of double refraction, but a more generalized law was needed to include all crystals.[21]

Brewster had developed a complete mathematical formulation of the laws of double refraction, based on the model of forces and axes. Even Huygens's work was used in Brewster's formulation; the physical explanation of undulations was rejected, but the mathematics was put to use. The laws that Brewster developed were mathematical descriptions and could be used without resort to a specific physical explanation. But it was clear that Brewster developed the laws as a result of his interest in forces, axes, and light particles. His continued search for the laws was based on his belief in the Newtonian model of forces and corpuscles. In proper context therefore, the mathematical laws and experimental descriptions were based on a specific physical explanation. Brewster realized that there were various problems which arose when certain questions were asked, such as: what produced the forces in the crystal and what were the positions of the axes with respect to the natural faces and planes of the crystal. "It is scarcely necessary to observe," he wrote, "that the phenomena of double refraction can not be referred to the ordinary action of attractive and repulsive forces."[22] Here was a whole new group of forces added to Newtonian optics to account for the phenomena. He dismissed the problem of the origin of these forces.

This new interpretation of double refraction and axes of force inside crystals did not mean a change in Brewster's explanation of forces outside the surface of bodies to account for reflection and refraction. The new polarization force took its place in addition to the older forces, within the Newtonian framework. "The force of double refraction and polarization extends without the surface of crystals, and within the sphere of the force which produces the partial reflection."[23] The light ray must first meet the zone of repulsion, before experiencing any polarization effects. If the ray was not reflected, the forces of the axes of the crystal took effect.

Brewster's development of a mathematical description and physical explanation for double refraction represents the completion of the Newtonian tradition of optics—a tradition that had been extended to include an explanation of all observed light phenomena. Brewster succeeded in including the old phenomenon of double refraction and the new phenomena of polarization in the Newtonian optical framework. But even though the explanatory power of the system seemed complete by 1820, it suffered from the same difficulties that it had had at the beginning of the century. The theory was very cumbersome in its application; it had some internal difficulties which were neglected; it did not explain how the variously sized particles of light could all have the same velocity, nor did it explain how those particles could be transmitted easily through dense, transparent substances. But, above all, the emission theory needed a new explanation for each new type of phenomenon discovered. It had to be patched and repaired to meet each new challenge, and it was only sufficient to explain the known phenomena, without the capability of predicting new observations. The whole structure had become ossified by 1820, but it took a full thirty years more for it to break apart and crumble away.

David Brewster was soon to be confronted by a new theory of optics. By 1820 Fresnel was beginning to win support for his new wave theory in France. Fresnel's development of an acceptable mathematical representation of a wave theory, based on the concept of transverse waves in the ether rather than longitudinal pulses or undulations, cleared the path for the eventual overthrow of the Newtonian emission theory.[24] The Newtonian theory could not withstand the double assault of a new explanation in terms of a transverse wave and the application of new and refined mathematics. Newtonian optics competed successfully with the old concept of undulations in the ether primarily because both were explanatory and there was no experimental evidence to provide a decisive choice. But the new suggestion of transverse waves lent itself startlingly well to mathematics. More important, the mathematics of the new wave theory predicted experimental results which were then observed in rapid succession.

David Brewster simply could not accept the notions of transverse waves in the ether. Brewster was therefore not greatly impressed by Fresnel's work, although he did recognize the elegance of the new mathematical developments. He found solace in his own experimental work and in his maintenance of what he believed to be the Newtonian tradition. His aversion to the ether was as much the result of his religious beliefs as it was a scientific objection. John Tyndall testified to this as follows:

In one of my latest conversations with Sir David Brewster he said to me that his chief objection to the undulatory theory of light was that he could not think the Creator guilty of so clumsy a contrivance as the filling of space with ether in order to produce light.[25]

Brewster remained firmly committed to the Newtonian theory of optics, even though the new wave theory was beginning to find favor in England. In the 1830s Brewster reverted to Newton's "Fits of Easy transmission and reflection" for the explanation of optical phenomena. In two papers in 1830, published in the *Philosophical Transactions of the Royal Society,* he referred back to his papers of 1813 through 1815, reasserting his position. Speaking of polarization by refraction and the law he developed in 1813, he made the remarkable comment: "During the sixteen years which have since elapsed, the subject does not seem to have made any progress."[26] Then, in speaking of elliptic polarization, Brewster wrote:

The theory of elliptic vibrations as given by Fresnel, will no doubt embrace the phenomena of elliptic polarization; and when the nature of metallic action shall be more thoroughly examined, we may expect to be able to trace the phenomenon under consideration to its true cause.

(I am familiar with M. Fresnel's formula only from the account given of it by Mr. Herschel.)[27]

Brewster was perfectly willing to admit that the mathematics developed by the wave theorists would describe the phenomena. But, he insisted that the "true cause" still lay in terms of particles and forces. Progress would be made only by the discovery of a better Newtonian explanation. Fresnel's work interested Brewster only because of the mathematics. Optical phenomena were still to be described in Newtonian terms.

In keeping with his commitment to Newtonian optics, Brewster wrote a popular textbook, *A Treatise on Optics,* in 1831. The textbook used Newton's *Opticks* as a basis and model. Most of Newton's descriptions were reproduced, with an account of the more recently discovered optical phenomena simply added on. Brewster attempted to separate the descriptions of nature from the attempts at explanation, in much the same way that Newton had presented the material in the *Opticks*. The reader was offered the important experiments and equations in optics, separated from any attempted physical explanation. The physical explanations were reserved for separate chapters within the book.

If the reader was not conscious of Brewster's adherence to the old Newtonian tradition from the context of the work, that fact was brought forcibly to his attention in a number of chapters. Brewster attempted to give as little credit as possible to the wave theory. When forced to refer to the accomplishments of this theory, he described it in the most unfavorable way possible. In a chapter entitled "On Fits of Reflexion and Transmission and of the Interference of Light," the concept of interference was presented as an afterthought—as a theory which was also capable of describing the phenomena. Brewster even preferred "Fits" to the new wave theory. An example will show Brewster's subtle methods of discounting the wave theory. His description of the ether might well have been paraphrased from Cotes' *Preface* to Newton's *Principia*:

In the undulatory theory, an exceedingly thin and elastic medium, called ether, is supposed to fill all space, and to occupy the intervals between the particles of all material bodies. The ether must be so extremely rare as to present no appreciable resistance to the planetary bodies which move freely through it.[28]

For David Brewster, maintenance of the Newtonian physical optics as late as 1831 in England was extremely conservative but not entirely unjustifiable. Brewster was perfectly willing to accept the new mathematical equations and experimental descriptions which were derived from the wave theory. But, at the same time, he was unwilling to give up forces and particles in favor of ether and waves. For Brewster, the system of optics involving material particles and forces acting at a distance was more compatible with his general view of nature than were ether and waves. Brewster had been responsible for the extension of Newtonian optics to include an explanation of all known optical phenomena in terms of particles and forces. It had become a cumbersome system, but it was fully capable of providing both mathematical descriptions and the physical explanations for optical phenomena. Brewster still preferred the Newtonian system of optics, with its long tradition and the prestige lent it by the name and reputation of Sir Isaac Newton.

Epilogue

While the choice between the Newtonian system of optics and the new wave theory was still a matter of personal preference in England in 1830, few young Englishmen became supporters of the old Newtonian tradition of optics. Young men, capable of understanding the mathematics of the wave theory, were won over by the simplicity and elegance of the new system. It was a complete system, with few assumptions, based solely on the behaviour of transverse waves in the ether. Where the Newtonian emission theory was forced to add a new explanation for every newly observed optical phenomenon, the wave theory enjoyed the secure position of being able to meet each new situation completely, as well as predicting the outcome of new experiments which it suggested. Where the Newtonian system had become cumbersome and unwieldy, the new wave theory by contrast seemed limited in application and scope only by the power of the mathematics which could be devised to advance it.[29] Some of the early supporters of the wave theory in England were William Hamilton, Humphrey Lloyd, George Airy, John Herschel, James MacCullagh, and James Challis. These men understood the mathematics of the new theory, as well as the physical optics. There were problems associated with the acceptance of an elastic ether, to be sure, but in general the mathematical elegance and predictive power of the wave theory carried the day.

The transition from Newtonian optics to the wave theory after 1830 is best illustrated by the works of Humphrey Lloyd of Dublin and of John Herschel. Both men began their work in optics as adherents of the corpuscular interpretation and both men became strong supporters of the wave theory in the 1830s. In 1834, the question of the choice between the old Newtonian system and the new wave theory was all but settled. Humphrey Lloyd presented a report to the British Association for the Advancement of Science which relegated the Newtonian system to the background.[30] This report was an extensive comparison of the accomplishments and premises of both the emission and the wave theories in all areas of physical optics. The account was comprehensive and unbiased: the wave theory, as it had developed since 1819, emerged as clearly superior in almost all areas of physical optics. The report made available a full account, in English, of the mathematical and experimental work performed in support of the wave theory in France for the 15 years preceding the report. The evidence was clear for all Englishmen to see.

The Newtonian system of optics did not suffer defeat in England in the 1830s, it simply failed to attract new adherents. The system had gone as far as its momentum would carry it. It had enjoyed great success in England, maintaining its dominance as the source of explanation in physical optics for more than a century. In the 1830s, it slipped into the background as the wave theory moved forward to replace it, spurred forward by its elegant mathematics and young and talented supporters.[31] The Newtonian system did not suffer outright defeat until the famous experiments on the velocity of light in different media, begun in Paris in 1849 by Jean Leon Foucault, were announced. Until this experimental proof of the error in the Newtonian system was established—an erroneous assumption that light would travel faster in denser media—the Newtonian system suffered neglect rather than defeat.

We have seen then the formulation, development, and culmination of a system of optics which captured and held men's allegiance for more than a century. It was a system of wide scope, admirably fitted to a universe envisioned in terms of forces and particles of matter. It was a universe, envisioned very much as Laplace described it, composed of central forces and bodies moving with completely describable motions. Both the Newtonian system of optics and the world view it suited so well found their culminations in the first quarter of the nineteenth century. Just as the wave theory was presented and accepted as a replacement for the old Newtonian structure of optics, so too a field theory was presented to rival the overall Newtonian world picture to which Newtonian optics was so well suited.

A system of explanation of the physical world, as comprehensive and as well founded upon experiment and observation as was the Newtonian system of optics, could never be denied all at once. It must be replaced by a new system of explanation, fundamentally different and more suitable in the sense that it lent itself more simply and directly to the new context of thoughts about the physical world. The Newtonian system and the world view of specific forces and material particles was too inflexible and cumbersome to deal with the new contexts which arose in the first half of the nineteenth century. The phenomena of elliptical and conical polarization, new observations on "invisible heat" and "dark rays," and the interaction and correlation of the multitude of physical forces—electrical, magnetic, chemical, and physiological—forced their attention upon the natural philosophers of the nineteenth century. The old Newtonian scheme had neither the conceptual flexibility nor the mathematical power to adapt itself to the emerging new world.

The development of the Newtonian system of optics was a stage in the history of man's attempt to understand the physical universe. It was a system which both described and explained, stimulated and satisfied. The wealth of experiment and observation associated with it presented descriptions of all optical phenomena. Its reliance upon zones of forces and particles of light succeeded in presenting explanations for these phenomena. During its period of acceptance, it sometimes stimulated men's investigation of the wide field of optics, and sometimes satiated basic curiosity about physical optics. It was a system capable of presentation in textbook form, offering satisfactory answers on most questions to beginner and savant alike.

The Newtonian system of optics stands as an example of man's ingenuity in creating a picture of the physical world, of his willingness to develop that picture according to his own preferences and inclinations, and of his tenacity maintaining it beyond its productive life span.

References

CHAPTER I

[1] For a discussion of this change of emphasis see A.E. Bell, *Newtonian Science*.

[2] There are many histories of the seventeenth century. Three good and readily available books are Sir George Clark, *The Seventeenth Century* (London: Oxford University Press, 1929; paper, 1969); Carl J. Friedrich, *The Age of the Baroque* (New York: Harper & Brothers, 1952); Maurice Ashley, *England in the Seventeenth Century,* The Pelican History of England No. 6 (Baltimore: Penguin Books, 1952).

[3] For an excellent treatment of the revival of ancient atomism in England see Robert Hugh Kargon, *Atomism in England From Hariot to Newton.* Hereafter cited as Kargon, *Atomism.*

[4] The following remarks on the revival of atomism are derived, in part from Kargon, *Atomism*, especially Chap. I.

[5] See R. Hooykaas, "Experimental Origin of Chemical Atomic and Molecular Theory Before Boyle," *Chymia* 2 (1949), 65-80.

[6] See Mary B. Hesse, *Forces and Fields,* a study of action at a distance in the history of physics, originally published 1961, especially Chaps. III and V. Hereafter cited as Hesse, *Forces and Fields.*

[7] Hesse, *Forces and Fields,* p. 72.

[8] See Hesse, *Forces and Fields,* p. 73, for a fuller presentation of this line of argument.

[9] Kargon, *Atomism,* p. 4.

[10] See Kargon, *Atomism,* for a full discussion of Harriot's atomism, especially Chap. III. The following discussion of Harriot is derived from Kargon.

[11] Kargon, *Atomism,* p. 27.

[12] See Kargon, *Atomism,* pp. 60-62. Kargon makes strong arguments for the influence of Pierre Gassendi on Hobbes' thought. He suggests that it was through the Newcastle Circle that the Epicurean atomism of Gassendi found its entry into England.

[13] Kargon, *Atomism,* p. 61.

[14] See Samuel Mintz, *Hunting of Leviathan* (Cambridge, 1962), pp. 40-41, quoted in Kargon, *Atomism,* p. 62.

[15] For Newton's religious concerns and his work as an historian, see Frank E. Manuel, *Isaac Newton, Historian* (Cambridge: The Belknap Press of Harvard University Press, 1963). Fuller treatment of Newton's views of God follow later in Chap. I.

[16] Kargon, *Atomism,* p. 106. The phrases taken from Bacon are found in Francis Bacon, *Works,* ed. J. Spedding, R. Ellis, and D. Heath, 15 vols. (Boston, 1860-64), Vol. VIII, pp. 59, 60, 114.

[17] T. Hobbes to Newcastle, July, 1636, in Historical Manuscripts Commission, *Portland Manuscripts* (London, 1891), II, 128. Quoted in Kargon, *Atomism,* p. 107.

[18] Descartes to Mersenne, 27 May 1638, in Norman Kemp Smith, *New Studies in the Philosophy of Descartes* (London, 1952), pp. 96-97. Quoted in Kargon, *Atomism,* p. 108.

[19] For an excellent discussion of the inappropriateness of thinking physical science can be objective, see E.A. Burtt, *The Metaphysical Foundations of Modern Science* (Doubleday Anchor Books). This was first published as *The Metaphysical Foundations of Modern Physical Science* in 1924, and revised for a second edition in 1932. Hereafter cited as Burtt, *Metaphyscial Foundations*.

[20] For an excellent treatment of the influence of Boyle, Barrow, and More, see Burtt, *Metaphysical Foundations*. The following presentation of Isaac Newton's science makes extensive use of Burtt's analysis.

[21] Henry More, *Immortality of the Soul*, contained in *A Collection of Several Philosophical Writings* (London, 1662), and *Enchiridion Metaphysicum* (London, 1671).

[22] Cited in Burtt, *Metaphysical Foundations*, p. 136. *Enchiridion Metaphysicum*, Chap. 9, Par. 21.

[23] Cited in Burtt, *Metaphysical Foundations*, p. 138. Boyle, *Works* (Birch edition), Vol. VI, p. 513, ff. Cf. *Divine Dialogues*, p. 16, ff.

[24] Cited in Burtt, *Metaphysical Foundations*, pp. 142-143. More, *Immortality of the Soul*, Preface.

[25] Burtt, *Metaphysical Foundations*, p. 143.

[26] See Frank Manuel, *Issac Newton, Historian* (Cambridge: The Belknap Press of Harvard University Press, 1963).

[27] General Scholium, Book III, *Isaac Newton's Mathematical Principles of Natural Philosophy and his System of the World*, Motte's Translation Revised by Florian Cajori (University of California Press, 1934), pp. 544-546. Hereafter cited as *Principia*.

[28] *Principia*, p. 547.

[29] *Principia*, pp. 529-530.

[30] *Principia*, p. 542. For mention of comet's tails, see *Principia*, pp. 494-495, 497, 521-526, 528-530, 542.

[31] For this interpretation of Newton's work, as well as the quote, see J.E. McGuire and P.M. Rattansi, "Newton and the 'Pipes of Pan'," *Notes and Records of the Royal Society of London*, Vol. 21, No. 2, December 1966. Hereafter cited as McGuire and Rattansi. For extension and support for this interpretation of Newton, see J.E. McGuire, "Body and Void and Newton's De Mundi Systemate: Some New Sources," and J.E. McGuire, "The Origin of Newton's Doctrine of Essential Qualities." The term "classical" scholia refers to a set of draft Scholia to Propositions IV to IX of Book III of the *Principia*.

[32] See McGuire and Rattansi for the elaboration of this excellent and appropriate historical presentation of Isaac Newton. The following will depend heavily upon their work.

[33] McGuire and Rattansi, p. 125.

[34] *Ibid.*

[35] *Ibid.*, p. 126.

[36] See E.A. Burtt, *Metaphysical Foundations*, pp. 297-302, for an extended treatment of this interpretation of Newton's theism.

[37] See *Principia*, p. xxxiii.

[38] See Burtt, *Metaphysical Foundations*, for an important and thought-provoking assessment of the naivete with which western civilization grasped the new science.

References

[39] Newton added to the number of queries in the various editions of the *Opticks*. The first English edition of 1704 contained the first 16 queries. The Latin version of this edition in 1706 contained six additional queries, including the now famous 31st Querie (which appeared as the 23rd query in the Latin edition). The second English edition, issued in 1718, contained the complete thirty-one queries.

[40] Sir Isaac Newton, *Opticks or a Treatise of the Reflections, Refractions, Inflections and Colours of Light* (Dover Publications, 1952) based on the Fourth Edition, London, 1730. The editions of the *Opticks* were as follows: *Opticks* (London: Smith and Welford, 1704); *Optice* (London: Smith and Welford, 1706); *Opticks* (London: W. & J. Innys, 1718), contained all 31 queries.

[41] See Kargon, *Atomism*, esp. pp. 133-135.

[42] See Hesse, *Forces and Fields*, esp. pp. 114-118.

[43] Henry Guerlac has drawn attention to the importance of the electrical experiments of Francis Hauksbee at the Royal Society to Newton's thoughts about the ether. See "Francis Hauksbee: experimentateur au profit de Newton," *Archives Internationale d'Histoire des Sciences* 16 (1963), 113-128; "Sir Isaac Newton and the Ingenious Mr. Hauksbee," *L'Aventure de la Science, Melanges Alexandre Koyré* I, ed. I.B. Cohen and René Taton (Paris: Herman, 1964); and "Newton's Optical Aether," *Notes and Records of the Royal Society of London* 22 (1967), 45-57.

[44] I have developed this argument at greater length in my article "Science As a Creative Art," contained in *Science, Technology and Culture*, ed. H.J. Steffens and H.N. Muller, (New York: AMS Press, Inc., 1974), pp. 85-133.

[45] Hesse, *Forces and Fields*, p. 74.

[46] For further elaboration of these general arguments see: Hesse, *Forces and Fields*, Chap. V, and E.A. Burtt, *Metaphysical Foundations*, Chaps. VII and VIII. This work is concerned with Newtonian optics, as distinct from Newton's optics. For Newton's own optics, see especially: Vasco Ronchi, *The Nature of Light, an Historical Survey*, trans V. Barocas, originally published as *Storia della Luce*, 1939; I. Bernard Cohen, *Franklin and Newton*; A.I. Sabra, *Theories of Light from Descartes to Newton*. There are, of course, a large number of articles on Newton's optics, easily located in the journals concerned with the history of science.

[47] Robert Smith, *A Compleat System of Opticks* in 4 Books, viz. a popular, a mathematical and a philosophical treatise (Cambridge: printed for the author, 1738). Hereafter cited as *Compleat System*.

[48] Smith probably used the third English edition of Sir Isaac Newton's *Opticks* (London: W. and J. Innys, 1721). The page references and quotations used by Smith in the *Compleat System* agree with those in this edition of the *Opticks*. Hereafter cited as *Opticks* (1721).

[49] For the most convenient source, see Roger Cotes, *Preface* to Sir Isaac Newton's *Mathematical Principles of Natural Philosophy and his System of the World*, Andrew Motte's translation revised by Florian Cajori (Berkeley: University of California Press, 1962). Hereafter cited as Cotes, *Preface*.

[50] See *Correspondence of Sir Isaac Newton and Professor Roger Cotes*, ed. Joseph Edelston (London, 1850). Facsimile reproduction (London: Frank Cass and Co., Ltd., 1969). Hereafter cited as Edelston, *Correspondence*.

[51] Edelston, *Correspondence*, pp. 152-153.

[52] Edelston, *Correspondence*, pp. 154-155.

[53] Edelston, *Correspondence,* letter 31 March 1713, p. 156.

[54] After his unpleasant experience of criticism of his first optical papers by Robert Hooke and others, Newton was unwilling to enter into a direct debate with Continental critics of his *Principia*. For an excellent treatment of Newton's critics in optics, see A.I. Sabra, *Theories of Light from Descartes to Newton*. For the Continental debate see H.G. Alexander, ed., *The Leibnitz-Clarke Correspondence* and Florian Cajori's Appendix 52 to Newton's *Principia* (University of California Press, 1962).

[55] Edelston, *Correspondence,* p. 159.

[56] Cotes, *Preface,* p. xxxi.

[57] See Alexandre Koyré, *Newtonian Studies* (Chicago: Phoenix Books, 1968), first published 1965, "The Significance of the Newtonian Synthesis," p. 16. This was a lecture given at the University of Chicago, November 3, 1948; published in *Archives Internationales d'Histoire des Sciences* 3 (1950), 291-311; reprinted in *Journal of General Education* 4 (1950), 256-268. Koyré supported this change of attitude with examples of a similar transition on the part of the French Newtonians, particularly Voltaire and Maupertuis.

[58] See George Huxley, "Roger Cotes and Natural Philosophy," for further details on Cotes. Also, *Dictionary of Scientific Biography* (New York: Scribners, 1971), Cotes, Roger, pp. 430-433 by J.M. Dubbey.

[59] Cotes, *Preface,* p. xxvii.

[60] See Alexander Koyré, "Attraction, Newton, and Cotes," *Archives Internationales d'Histoire des Sciences* 14 (1961), 225-236, included in *Newtonian Studies* (Phoenix Books, 1965), pp. 273-282. Koyré provides an excellent discussion of Cotes and attraction.

[61] J. Edelston, *Correspondence of Sir Isaac Newton and Professor Cotes* (London, 1850).

[62] Smith was responsible for the publication of *Harmonia Mensurarum sive Analysis et Synthesis per Rationum et Angulorum Mensuras promotae; accendunt alia opuscula mathematica per Rogerum Cotesium. Edidit et Auxit Robertus Smith* (Cambridge, 1722) and *Hydrostatical and Pneumatical Lectures* by Roger Cotes, the second edition by Robert Smith (Cambridge, 1747). Smith dedicated these lectures to William, Duke of Cumberland.

[63] Smith, *Compleat System,* p. 1.

[64] *Ibid.,* p. 2.

[65] Alexandre Koyré, "Concept and Experience in Newton's Scientific Thought," in *Newtonian Studies* (Chicago: University of Chicago Press, 1968), p. 43. (Originally published as "L'Hypothese et l'expérience chez Newton," *Bulletin de la Société Française de Philosophie* 50 (1956), 59-79.) See Koyré for a discussion of Newton's hesitancy.

[66] Newton, *Principia,* p. 226.

[67] *Ibid.,* pp. 230-231. Proposition XCVII Problem LXVII: "Supposing the sine of incidence upon any surface to be in a given ratio to the sine of emergence; and that the inflection of the paths of those bodies near that surface is performed in a very short space, which may be considered as a point, it is required to determine such a surface as may cause all the corpuscles issuing from any one given place to converge to another given place."

[68] *Opticks* (1721), Book II, partIII, prop. 8, p. 266.

[69] Smith, *Compleat System,* p. 87.

References

[70] *Ibid.*, p. 91.

[71] *Ibid.*, Figures 288-291.

[72] *Ibid.*, p. 93.

[73] For a very interesting presentation of the role of textbooks in the stages of development of science, see Thomas S. Kuhn, *The Structure of Scientific Revolutions*, Second Edition, Enlarged, *International Encyclopedia of Unified Science* Vol. 2, No. 2 (Chicago: University of Chicago Press, 1970).

[74] *Opticks* (1721), p. 300.

[75] Smith, *Compleat System*, p. 85.

[76] *Ibid.*, p. 90.

[77] *Opticks* (1721), p. 245.

[78] See W.B. Hardy, "Historical Notes upon Surface Energy and Forces of Short Range."

[79] Francis Hauksbee, *Physico-Mechanical Experiments* (London, 1709).

[80] Quoted in W.B. Hardy, p. 375.

[81] See *ibid.*, for further discussion of these forces.

[82] See Richard Westfall, "The Development of Newton's Theory of Color."

[83] See I. Bernard Cohen, *Isaac Newton's Papers and Letters on Natural Philosophy*, pp. 27-240, for Newton's early optical papers.

[84] *Compleat System*, p. 92; *Opticks* (1721), p. 347, Query 29.

[85] Smith, *Compleat System*, p. 97.

[86] See A.I. Sabra, *Theories of Light from Descartes to Newton* for an excellent discussion of Newton's "fits."

[87] Smith, *Compleat System*, p. 93.

[88] *Opticks* (1721), Prop, XII, Book II, p. 253.

[89] Qu. 5—"Do not Bodies and Light act mutually upon one another; that is to say, Bodies upon light in emitting, reflecting, refracting, and inflecting it and Light upon Bodies for heating them, and putting their parts into a vibrating motion where in heat consists . . .?"

[90] Edmund Whittaker laments this situation in *A History of the Theories of Aether and Electricity* Vol. I, p. 31.

[91] See Robert E. Schofield, *Mechanism and Materialism*, for an excellent treatment of Newtonian science in the eighteenth century; especially concerning theories of matter. Part I, "Mechanism and Dynamic Corpuscularity, 1687-1740" is an excellent supplement to this volume in that it treats the broader questions involving general theories of matter.

[92] Professor Schofield has given a different interpretation of Smith's mention of the aether. See pages 32-34.

[93] (London, 1704).

[94] *Dioptrica Nova*, a treatise of Dioptricke in two parts wherein the various effects and appearances of spherick glasses both convex and concave, single and combined, in telescopes and microscopes, together with their usefulness in many concerns of humane life, are explained. (London, 1692-4); Second edition (London, 1709).

⁹⁵(London: S. Palmer, 1728).

⁹⁶Pemberton, p. 318.

⁹⁷Jean Théophile Desaguliers, *A Course of Experimental Philosophy,* 2 vol. (London, 1734-44). See I.B. Cohen, *Franklin and Newton,* for a discussion of Desaguliers as a Newtonian, especially pp. 243-261.

⁹⁸(London: James Hodges, 1740).

⁹⁹Benjamin Martin, *A New and Compendious System of Optics,* Preface, xiii. The references are to William Molyneux (1656-1698), *Dioptrica Nova* (London, 1692), 2nd edition, 1709; and David Gregory (1661-1708), *Elements of catoptrics and dioptrics,* Trans. W. Browne (London, 1735). Original Latin edition (Oxon: 1695).

¹⁰⁰Martin, *Philosophical Grammar,* p. 57. On the topic Martin wrote: "The Newtonians very justly make Primigenial Light to consist in a certain Motion of the Particles of luminous Bodies, which force out and off the Said Bodies certain exceeding and inconceivably small particles, which, with incredible Force, are every Way emitted in straight Lines: And Derivative Light to consist not in an Endeavour to Motion, but on a real Motion of those Particles emitted from Bodies, as aforsaid."

¹⁰¹*Ibid.,* p 70.

¹⁰²Benjamin Martin, *New Elements of Optics* or the Theory of the Aberrations, Dissipation, and Colours of Light: of the General and Specific Refractive Powers of Densities of Mediums; The Properties of Single and Compound lenses: and the Nature, Construction, and Use of Refracting and Reflecting Telescopes and Microscopes of every sort hitherto published (London, 1759).

¹⁰³Benjamin Martin, *New Elements of Optics,* pp. 56-57.

¹⁰⁴Richard Helsham, *A Course of Lectures in Natural Philosophy,* Professor of Physick and Natural Philosophy in the University of Dublin. T. Rutherford, *A System of Natural Philosophy being a course of lectures in Mechanics, Optics, Hydrostatics and Astronomy* Which are read in St. Johns College, Cambridge, 2 Vol. John Rowning, *A Compendious System of Natural Philosophy* in Four Parts (London, 1737-43). The work had gone through six editions by the London edition of 1767.

¹⁰⁵See Robert E. Schofield, *Mechanism and Materialism* for an excellent discussion of the transition about 1740. See also I.B. Cohen, *Franklin and Newton* for a discussion of the importance of the *Opticks* to the development of experimental science in the eighteenth century.

¹⁰⁶The attempts at using the ether in optics will be discussed in greater detail in Chaps. II and III.

¹⁰⁷It has been suggested that Newtonian optics reached its limits in the eighteenth century because of its reliance upon mechanical, corpuscular explanations. See Peter Anton Pav, *Eighteenth Century Optics: The Age of Unenlightenment* (Unpublished Ph.D. dissertation, Indiana University, 1964). Available from University Microfilms, Inc., Ann Arbor, Michigan #65-3510. I disagree with this interpretation because it gives what I consider to be an incorrect connotation to the condition of optics in the eighteenth century. Newtonian optics was a functioning system which gave reasonable assurance of being capable of extension to meet new problems in optics. It proved capable of meeting challenges in the eighteenth

References

century (see my Chap. II) and it was capable of rapid and convincing extension in the early nineteenth century to include the new phenomena of polarization, as we shall see in Chap. IV. Men adhered to Newtonian optics in the eighteenth century for good reason and with the confidence of extensive experimental and observational support. There was as much reason to maintain Newtonian optics in the eighteenth century as there was to maintain any organized system of scientific ideas. The argument that it had reached its limit implies that had it not been maintained, new developments would have occurred. This does not seem to me to be an appropriate line of reasoning, considering both the scientific context of the times and what occurred later in the development of Newtonian optics. See Pav's work for a treatment of Newtonian optics in France.

References

CHAPTER II

[1] "A letter from Mr. John Dolland to James Short, A.M.F.R.S. concerning a Mistake in M. Euler's Theorem for correcting the Aberrations in the Object Glasses of refracting Telescopes," *Phil. Trans.* 48 (1752), 289. For Euler's paper see "Sur la perfection des verres objectifs des lunettes," Akademie der Wissenschaften, Berlin. *Histoire . . . avec les memoires,* 1747, pp. 274-296, with tables.

[2] Dolland wrote as criticism: "This gentleman puts $m:1$ for the ratio of the refraction out of air into glass of the mean refrangible rays, and $M:1$ for that of the least refrangible. Also for the ratio of refraction out of air into water of the mean refrangible rays he puts $n:1$, and for the least refrangible $N:1$. As to the numbers, he makes $m = 31/20$, $M = 77/50$ and $n = 4/3$: which so far answer well enough to experiments. But the difficulty consists in finding the value of N in a true proportion to the rest.

Here the author introduced the supposition above mention'd; which is, that m is the same power of M, as n is of N; and therefore puts $n = m^a$, and $N = M^a$. Whereas, by all the experiments that have hitherto been made, the proportion will come out thus $M-1:n-1::m-M:n-N$." *Phil. Trans.* 48 (1752), 290.

[3] Klingenstierna, Samuel, *Dictionary of National Biography,* Vol. 5, p. 1101.

[4] "An Account of some Experiments Concerning the different Refrangibility of Light" by John Dolland.

[5] *Ibid.*, p. 736.

[6] *Ibid.*, p. 740.

[7] The case of Chester Moor Hall will not be considered here, but is subject to my further investigation. Hall apparently succeeded in constructing achromatic lenses in 1733. The clue for his search for such a lens seems to have come from a study of the human eye. He reasoned that since nature produced such a lens, perhaps it was possible for man. Hall never made his success known to the public.

[8] Joseph Priestley, *The History and Present State of Discoveries relating to Vision, Light and Colours,* p. 474.

[9] "Rules and Examples for limiting the Cases in which the Rays of refracted Light may be reunited into a colourless Pencil." In a letter from P. Murdock M.A. ad F.R.S. to Robert Symmer, Esq.: F.R.S. January 3, 1763.

[10] Frank E. Manuel, "Newton as Autocrat of Science." The reference is to Francesco Algarotti, *Sir Isaac Newton's philosophy explain'd for the use of the ladies,* trans. Elizabeth Carter, 2 vol. (London, 1739). Original Italian version 1737.

[11] See Marjorie Hope Nicolson, *Newton Demands the Muse, Newton's Opticks and the 18th Century Poets,* p. 11. This work treats many facets of Newton's influence in the eighteenth century.

[12] *Ibid.*, p. 92.

[13] "A Letter from Mr. T. Melvil to the Rev. James Bradley, D.D.F.R.S. With a Discourse Concerning the Cause of the different Refrangibility of the Rays of Light."

[14] *Ibid.*, pp. 263-264.

[15] *Ibid.*, p. 267.

[16] These papers were published by the Royal Philosophical Society of Edinburgh as Article IV, "Observations on Light and Colours," in *Essays and Observations Physical and Literary* (Edinburgh, 1770-71).

[17] *Ibid.*, pp. 50-51.

[18] Henry Pemberton, *A View of Sir Isaac Newton's Philosophy*, p. 376.

[19] *Phil. Trans.* 48 (1753), 269.

[20] Joseph Priestley, *The History and Present State of Discoveries relating to vision, light and colours*, pp. 404-405.

[21] *Ibid.*, p. 383.

[22] *Ibid.*, p. 385.

[23] See Fontenelle's description of Homberg's work in *Histoire de l'Academie Royale des Sciences avec les memoires de mathematique et physique*, H 21, 1708. See Pav for a more extended treatment of the French experiments on the momentum of light, especially pp. 76-84.

[24] See J.J. Dortous de Mairan, "Eclaircissemens sur le traite physique et historique de l'aurore boreale, qui fait la siut des memoires de l'Academie Royale des Sciences, annee 1731," *Histoire de l'Academie Royale des Sciences avec les memoires de mathematique et physique*, M 427, 1747. Quote taken from Pav, p. 79.

[25] Robert E. Schofield, "Joseph Priestley, Natural Philosopher," p. 5.

[26] See Robert E. Schofield, *Mechanism and Materialism*, p. 96. See especially Part II, "Aether and Materialism, 1740-1789," for a discussion of the transition from mechanism to materialism and for the revival of speculations on the aether after 1740.

[27] See Russell McCormmach, "John Michell and Henry Cavendish: Weighing the Stars," *British Journal for the History of Science* 4 (1968), 130.

[28] See Robert Schofield, *Mechanism and Materialism*, especially Part III.

[29] See Robert E. Schofield, ed., *A Scientific Autobiography of Joseph Priestley (1733-1804)* (Cambridge: The M.I.T. Press, 1966), Nos. 39 and 40.

[30] Priestley, *Vision, Light and Colours*, pp. 387-388.

[31] *Ibid.*, pp. 389-390.

[32] See Lancelot Law Whyte, *Roger Joseph Boscovich (1711-1787)*.

[33] First published in Vienna in 1758 and 1759, it was revised by Boscovich for a second edition published in Venice in 1763, and reprinted in Venice in 1764 and 1765. Conveniently available in a Latin-English edition, *A Theory of Natural Philosophy* (Chicago: Open Court, 1922) which is a reprint of the Venetian edition of 1763. Hereafter cited as Boscovich, *Theory*.

[34] Priestley, *Vision, Light and Colours*, pp. 390-391. (For Boscovich's treatment of light, see his own summary, Boscovich, *Theory*, pp. 166-177.)

[35] See *ibid.*, pp. 306-310 for description of "Fits" and Priestley's negative attitude toward them.

[36] *Ibid.*, pp. 310-311.

References

[37] *Ibid.*, p. 330.

[38] *Ibid.*, p. 310.

[39] Priestley suggested the source of Michell's ideas was Baxter's *On the Immateriality of the Soul.* See Priestley, pp. 391-392. Robert E. Schofield has suggested that the more possible and direct source was Rowning's work, used by Priestley, and a text at Cambridge when Michell was a student there. See Schofield, *Mechanism and Materialism,* footnote 11, p. 245. Schofield implies that Michell's scheme was Boscovichian. This may be so, or it may be equally possible that Michell developed his explanation for the colors of thin plates by an extension of the use of zones of attractive and repulsive force, a major characteristic of Newtonian optics.

[40] *Phil. Trans.* 57 (1767), 234-264.

[41] *Ibid.*, pp. 262-264.

[42] *Phil. Trans.* 74 (1784), 35-57. The paper had the long title "On the means of discovering the distance, magnitude, etc. of the fixed stars in consequence of the diminution of the velocity of their light, in case such a diminution should be found to take place in any of them, and such other data should be procured from observations, as would be farther necessary for that purpose." See Russell McCormmach, "John Michell and Henry Cavendish: Weighing the Stars," pp. 126-155, for a discussion of the 1767 paper and especially the 1784 paper. McCormmach is concerned primarily with astronomy, but his article clearly indicates the connection between Michell's assumptions in optics and his suggestions in sidereal astronomy.

[43] See letter of Michell to Cavendish, 26 May 1783 in *Phil. Trans.* 74 (1784), 35-36. Michell's calculations appear in Priestley, pp. 786-791.

[44] *Principia,* Book I, p. 125. Proposition XXXIX, Problem XXVII: "Supposing a centripetal force of any kind, and granting the quadratures of curvilinear figures; it is required to find the velocity of a body, ascending or descending in a right line, in the several places through which it passes, as also the time in which it will arrive at any place, and conversely."

[45] Priestley, pp. 789-790.

[46] See McCormmach, p. 142. The quote is from a letter of Michell to Cavendish, 2 July 1783, Devonshire Collections, Chatsworth. See McCormmach also for a clear discussion of the astronomical problems and assumptions involved in Michell's paper.

[47] Michell, *Phil. Trans.* 74 (1784), 51. See Newton's *Opticks,* Book I, Prop. VI, Theor. V, p. 79.

[48] Newton, *Opticks,* pp. 81-82.

[49] This account has relied heavily upon McCormmach's paper, and his presentation of quotations from letters between Michell and Cavendish in the Devonshire Collections, Chatsworth.

[50] Cavendish to Michell, 27 May 1783, Devonshire Collections, Chatsworth. Presented in McCormmach, p. 148.

[51] Newton, *Principia,* p. xviii.

[52] Russell McCormmach, "Henry Cavendish: A Study of Rational Empiricism in Eighteenth-Century Natural Philosophy," p. 295.

[53] See *The Scientific Papers of the Honorable Henry Cavendish,* F.R.S., Vol. II: Chemical and Dynamical, ed. E. Thorpe (Cambridge, 1921), p. 437. Suggested in footnote 77, McCormmach, "John Michell and Henry Cavendish," p. 143.

⁵⁴Herschel wrote this in February, 1780. See "On the Central Powers of the Particles of Matter," published in *The Scientific Papers of Sir William Herschel*, 2 Vols. (London, 1912), I, p. lxxviii.

⁵⁵*Ibid.*, p. lxx.

⁵⁶*Ibid.*, p. lxxiv.

⁵⁷Robert E. Schofield has argued persualsivley that the whole of Sir William Herschel's career in astronomy should be analyzed in terms of his persistent belief in the frame of reference of Newtonian central forces. This would modify the standard image of "Herschel the neo-Baconian observer" considerably. Schofield suggests that perhaps Herschel was similar to Michael Faraday in that he had a definite theoretical framework, but chose not to reveal it explicitly in his published papers. See Schofield, *Mechanism and Materialism*, pp. 249-254. Herschel's orientation toward Newtonian optics supports this interpretation.

⁵⁸The paper was read by Playfair on April 7, 1788, and was entitled "On the Motion of Light, as affected by refracting and reflecting Substances, which are also in Motion." Published in the *Transactions of the Royal Society of Edinburgh* 2 (1788), 83-111.

⁵⁹*Ibid.*, p. 96.

⁶⁰*Ibid.*, pp. 97-98.

⁶¹*Ibid.*, p. 106.

⁶²The interpretation of Robison's treatment of optics presented here differs from that presented in G.N. Cantor, "Henry Brougham and the Scottish Methodological Tradition," *Studies in History and Philosophy of Science* 2, No. 1 (1971), 77.

⁶³See Henry Brougham, "Experiments and Observations on the Inflection, Reflection, and Colours of Light," *Phil. Trans.* 86 (1796), 227-277; "Farther Experiments and Observations on the Affections and Properties of Light," *Phil. Trans.* 87 (1797), 352-385. These were reprinted in *A Journal of Natural Philosophy, Chmeistry and the Arts by William Nicholson* I (1797) and II (1798).

⁶⁴*Phil. Trans.* 86 (1796), 235. Brougham's reference to Newton's explanation of reflection was to *Opticks* (1721), Bk. II, Part III, Prop. 8. The theory of Boscovich refers to the *Nova Theoria Philosophie Naturalis*.. Brougham saw Boscovich's use of forces of attraction and repulsion as support, in a very general way, for his procedures in optics.

⁶⁵*Ibid.*, pp. 266-267.

⁶⁶*Ibid.*, p. 271.

⁶⁷*Ibid.*, p. 227.

⁶⁸*Ibid.*. p. 233.

⁶⁹*Ibid.*, p. 249.

⁷⁰*Ibid.*, pp. 249-250. The actual calculation is of interest as an example of Brougham's method. (See Fig. 10.)

"In fig. 8 let EC be the reflecting surface, DH the perpendicular, as AB a ray incident at B, and produced to F, and reflected to GB, draw GH parallel to FB, and GF to HB. Then HB : (HG:) : : sin HGB: sin HBG, or : : sin GBF: sin HBG. But GBF is the supplement of GBA, the sum of the angles of reflection and incidence; therefore $HB : BF$: : the sine of the sum of the

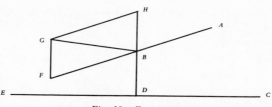

Fig. 10   Forces on light corpuscles.

angles of reflection and incidence, to the sine of the sum of the angles of reflection; so that if I be the angle of incidence, R that of reflection, V the velocity of light, and F the incidence, R that of reflection V the velocity of light and F the reflection force:

$$F = \frac{V \times \sin(R + I)}{\sin R}$$

By accommodating the formula of the different cases, we obtain F in all the rays; and the ratio of F in one set to I in another being required, we have (by striking out V, which is constant)

$$F:F' :: \frac{\sin(R + I)}{\sin R} : \frac{\sin(R' + I')}{\sin R'}$$

Suppose we would know F and F' in the red and violet respectively: $I = 50°48' - R = 50°21'$ and $I' - 51°15'$; then

$$F:F' :: \frac{\sin 101°9'}{\sin 50°21'} : \frac{\sin 102°3'}{\sin 51°15'}$$

Performing the division in each by logarithms, and finding the natural sines corresponding to the quotients; $F = F' :: 1275:1253.$"

[71] *Ibid.*, pp. 250-251.

[72] Brougham's second paper, *Phil. Trans.* 87 (1797), 352-385. Reprinted in *Nicholson Journal* II (1798), 151.

[73] *Ibid.*, p. 196.

[74] *Ibid.*, p. 192.

[75] For an excellent treatment of the early undulatory theorists see A.I. Sabra, *Theories of Light from Descartes to Newton*. This work should be consulted for Descartes's optical works, Fermat's improvements of Cartesian optics, Huygens's theory and his special kind of Cartesianism, Newton's early optical works, and the controversy which arose as a result of Newton's first papers in optics. See also my review of Sabra's book in *Brit. J. Phil. Sci.* 22 (1971), 55-57; the book is an outstanding treatment of optics before the Newtonians.

[76] See also Margaret 'Espinasse, *Robert Hooke* (London: Heinemann Ltd., 1956), and Vasco Ronchi, *The Nature of Light* for discussion of Descartes, Grimaldi, and Hooke, especially Chaps. 4 and 5.

[77] Robert Hooke, *Micrographia or Some physiological Descriptions of Minute Bodies made by Magnifying Glasses with Observations and Inquiries thereupon* (London: Royal Society, 1665), p. 54. Easily

available today in facsimile reproductions. The Dover edition (1961) was used here. Hereafter cited as *Micrographia*.

[78] *Ibid.*, p. 56-57.

[79] *Ibid.*

[80] *Ibid.*, p. 96.

[81] *Ibid.*, p. 60.

[82] *Ibid.*, p. 64.

[83] Huygens did not publish the Treatise until 1690, in Leyden, because he hoped to translate it into Latin. But after a delay of twelve years, he decided to publish the work under the title: "Traite de la lumiere ou sont expliquees les causes de ce qui luy arrive dans la reflexion et dans la refraction. Et particulierement dans L'etrange refraction du cristal d'Islande." par C.H.D.Z.) Christian Huygens de Zuylichem).

[84] Christian Huygens, *Treatise on Light,* translator Silvanis P. Thompson (London: Macmillan Co., 1912). The edition used here was published in Chicago, Encyclopedia Britannica Great Books, 1955, p. 554. Hereafter cited as *Treatise on Light.*

[85] *Ibid.*, p. 557.

[86] *Ibid.*, pp. 557-558.

[87] *Ibid.*, p. 560.

[88] *Ibid.*, p. 561.

[89] *Principia*, p. 382.

[90] *Treatise on Light,* p. 561.

[91] *Ibid.*, p. 562. (See Fig. 11.) "Thus If DCF is a wave emanating from the luminous point A, which is its centre, the particle B, one of those comprised within the sphere DCF, will have made its particular or partial wave KCL, which will touch the wave DCF at C at the same moment that the principal wave emanating from the point A has arrived at DCF; and it is clear that it will be only the region C of the wave KCL which will touch the wave DCF, to wit, that which is in the straight line drawn through AB. Similarly the other particles of the sphere DCF, such as bb, dd, etc., will each make its own wave. But each of these waves can be infinitely feeble only as compared with the wave DCF to the composition of which all the others contribute by the part of their surface which is most distant from the center A."

[92] *Ibid.*, p. 562.

[93] See *Treatise on Light,* Chaps. II-VI.

[94] See Vasco Ronchi, *The Nature of Light,* p. 202, for a discussion of Huygens's optics.

[95] See *Opticks* (4th edition, 1730, New York: Dover, 1952), p. 266.

[96] *Treatise on LIght,* p. 566. "But the thing to be remarked in our demonstration is that it does not require that the reflecting surface should be considered as a uniform plane, as has been supposed by all those who have tried to explain the effects of reflection; but only an evenness such as may be attained by the particles of the matter of the reflecting body being set nearer to one another; which particles are larger than those of the ethereal matter. . . ."

[97] See A.W. Badcock, "Physical Optics at the Royal Society," for a discussion of papers on optics presented to the Royal Society.

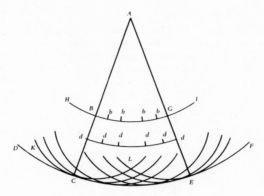

Fig. 11 Spherical wavefront.

[98] See H.G. Alexander (ed.), *The Leibnitz-Clarke Correspondence.*

[99] For an excellent discussion of these objections see Alexandre Koyré, "Newton and Descartes," in *Newtonian Studies,* pp. 53-200. See also Leonard M. Marsak, "Cartesianism in Fontenelle and French Science, 1686-1752," for the specific case of Fontenelle's difficulties with Newton's work.

[100] See Mary Hesse, *Forces and Fields,* for a discussion of one facet of this problem, action-at-a-distance; especially Chap. VII.

[101] Peter Anton Pav has a discussion of the reception of Newtonian optics in France in his "Eighteenth-Century Optics: The Age of Unenlightenment."

[102] It has been suggested that a concept of frequency related to color could be derived from Newton's work on the colors of thin films and his suggested "Fits of Easy Transmission and Reflection." While it is true that Newton did suggest a periodic disturbance in the ether to explain the regular production of colors, this suggestion was certainly a special case to be applied only to the puzzling phenomena of thin films, soap bubbles, and "Newton's" rings. That suggestion was not meant to apply to the other observed color phenomena, such as refraction. Certainly Newton's followers restricted this suggestion to special cases. As we have seen, the Newtonian system met the problem of thin films, as best it could, by attempting to explain away the problem by stressing the thickness necessary for the colors to be produced. Newton's concept of "Fits" was seldom used and generally avoided by the English. It was a rather embarrassing aspect of optics for the Newtonians. For Malebranche's works on colors, see "Reflexions sur la lumiere et les couleurs, et la generation du feu," par Père Malebranche, *Histoire de l'Academie Royale des Sciences.* Année MDCXCIX, A Paris, MDCCXXXII, pp. 22-36. See also Pierre Duhem, "L'Optique de Malebranche," *Revue de Metaphysique et de Morale* 23 (1916), 37-91.

[103] "Essai d'une explication physique des couleurs engendrees sur des surfaces extrement minces" and "Recherches physiques sur la diverse refrangibilite des rayons de lumiere."

[104] *Lettres a une princesse d'Allemagne sur quelques sujets de physique et de philosophie* (Petersburg, 1768-73, in 3 parts). The edition used here was *Letters of Euler on different subjects in Physics and Philosophy addressed to a German Princess,* trans. Henry Hunter (London, 1802).

[105] *Ibid.*, p. 112.

[106] See Robert E. Schofield, *Mechanism and Materialism,* Part II, pp. 89-231, for a broad discussion of the use of imponderable fluids and the uses of the ether, 1740-1789. See also I. Bernard Cohen, *Franklin and Newton,* especially for Franklin's objections to the corpuscular concept of light, and R.H. Silliman, *Augustin Fresnel and the Establishment of the Wave Theory of Light,* Chap. II, for a discussion of the undulatory theory in France during the eighteenth century. Hereafter cited as Silliman, *Fresnel.*

[107] Brian Higgins, *A Philosophical Essay Concerning Light.*

[108] *Ibid.*, p. 251.

[109] *Ibid.*, p. 253.

[110] *Ibid.*, p. 256.

References

CHAPTER III

[1] It has been suggested that Young's work on the undulatory theory was inspired by Newton's *Opticks*. See I. Bernard Cohen, *Preface* to the Dover edition of Newton's *Opticks* (New York: Dover Publications, 1952), pp. xxxix-xlvii. There Professor Cohen writes: "Young was indebted to Newton for more than the data for computing wave-lengths, wave-numbers, and frequencies; the whole wave theory of light was developed by him from the suggestions in Newton's *Opticks*, with several important additions, chiefly (1) considering the waves in the aether to be transverse, (2) supplementing the wave theory by the principle of interference." This view, while plausible, should be modified after a careful study of Young's educational development and especially his early papers. My work presents quite a different interpretaion of Young's development of the undulatory theory. This position should become clear as the chapter unfolds.

[2] Hudson Gurney, *Memoir of the Life of Thomas Young*, p. 9.

[3] The details of Young's choice of the medical profession as well as the general biographical information can be found in two biographies; George Peacock, *Life of Thomas Young* and Alexander Wood, *Thomas Young, Natural Philosopher*. Wood's work is readily available so limited biographical information will be included in this work.

[4] Young's personal bibliography written in his own hand is reproduced in Hudson Gurney's *Memoir of the Life of Thomas Young*.

[5] Alexander Wood, *Thomas Young, Natural Philosopher*, p. 49. Hereafter cited as Wood.

[6] Andrew Dalzel, *History of the University of Edinburgh from its first Foundation*, with a memoir of the Author by C. Innes. Hereafter cited as Dalzel.

[7] Letter dated July 5, 1797. Dalzel, pp. 143-144.

[8] Wood, p. 65.

[9] Letter dated Emmanuel College, Cambridge, July 8, 1798. Dalzel, p. 161.

[10] *Phil. Trans.* 90 (1800), 106..

[11] *Ibid.*

[12] *Iid.*, p. 118.

[13] *Ibid.*, p. 114.

[14] *Ibid.*, pp. 125-126.

[15] *Ibid.*, p. 128. Geoffrey Cantor has provided a detailed study of Young's use of the ether in his article "The Changing Role of Young's Ether," *The British Journal for the History of Science* 5, No. 17 (1970), 44-62. Cantor's article treats Young's optical writings from the perspective of Young's use of the ether as a possible foundation for a system of natural philosophy. He believes the ether was an important part of Young's early scientific work and concluded that Young's "rejection of the ether distribution hypothesis was paralleled by a change in his role as a scientist. He changed from a natural philosopher, attempting to construct a universal philosophy of nature, to a specialist tackling specific problems." I did not find Young's

conception of the ether to be as important to his optical work as Cantor suggests. I found Young to be primarily concerned with establishing his concept of interference. His readiness to abandon the ether density concept, which Cantor describes, suggests to me its lack of central importance. An undulatory theory requires an ether of some kind and Young had at least to mention the possibility of explanation in terms of that ether. He quickly abandoned ethereal explanations when he realized they were untenable, to concentrate upon experimental and observational support for his theory of interference.

[16] *Phil. Trans.* 90 (1800), 128.

[17] *Ibid.*, pp. 130–131.

[18] *A Journal of Natural Philosophy, Chemistry, and the Arts by William Nicholson* (August 1801), 1. Hereafter cited as *Nicholson Journal.*.

[19] Letter to Dalzel, dated 48, Wellbeck Street, June 27, 1801. Dalzel, p. 206.

[20] Young's reply in 1804 to criticism by Brougham found in John Tyndall, *Six Lectures on Light*, pp. 241–242.

[21] *Phil. Trans.* 92 (1802). The reference is to Hooke's *Micrographia*, pp. 65–67.

[22] Edmund Halley, *Phil. Trans.* 14 (1684), 681. Newton, *Principia*, Prop. XXIV, Theorem XIX, p. 435.

[23] See I. Bernard Cohen, "The First Explanation of Interference." Professor Cohen gives a clear discussion of the problem of tides, and Newton's solution to them, pp. 105–106.

[24] Newton's argument went as follows: "Further, it may happen that the tide may be propagated from the ocean through different channels towards the same port, and may pass quicker through some channels than through others; in which case the same tide, divided into two or more succeeding one another, may compoint new motions of different kinds. Let us suppose two equal tides flowing towards the same port from different places, one preceding the other by six hours; and suppose the first tide to happen at the third hour of the approach of the moon to the meridian of the port. If the moon at the time of the approach to the meridian was in the equator, every six hours alternately there would arise equal floods, which, meeting with as many equal ebbs, would so balance each other that for that day the water would stagnate and be quiet." *Principia*, p. 439.

[25] The Bakerian Lecture "On the Theory of Light and Colours," *Phil. Trans.* 92 (1802).

[26] *Nicholson Journal* (1801), 163.

[27] *Phil. Trans.* (1800), 143–144.

[28] *Nicholson Journal* (1801), 164.

[29] *Nicholson Journal* (May–August 1802), 264; (September–December 1802), 39; (September–December 1802), 145. John Gough (1757–1825) mathematician and natural philosopher, teacher of Dalton and Whewell.

[30] *Ibid.* (September–December 1802), 39.

[31] Robert Smith, *Harmonics, or the philosophy of musical sounds, 2nd edition much improved and augmented.*

[32] *Nicholson Journal* (September–December 1802), 145

[33] *Phil. Trans.* (1802), 13.

References

[34] *Phil. Trans.* (1802), 12.

[35] *Ibid.*, p. 29.

[36] *Ibid.*, 1802, p. 34. Proposition VIII:

"When two Undulations, from different Origins, coincide either perfectly or very nearly in Direction, their joint effect in a Combination of the Motions belonging to each.

"Since every particle of the medium is affected by each undulation, wherever the directions coincide, the undulations can proceed no otherwise than by uniting their motions, so that the joint motion may be the sum or difference of the separate motions, accordingly as similar or dissimilar parts of the undulations are coincident.

"I have, on a former occasion, insisted at large on the application of this principle to harmonics; and it will appear to be of still more utility in explaining the phenomena of colours. The undulations which are now to be compared are those of equal frequency. When the two series coincide exactly in point of time, it is obvious that the united velocity of the particular motions must be greatest, and, in effect at least, double the separate velocities; and also, that it must be smallest, and if the undulations are of equal strength, totally destroyed, when the time of the greatest direct motion of belonging to one undulation coincides with that of the greatest retrograde motion of the other. In intermediate states, the joint undulations will be of intermediate strength, but by what laws this intermediate strength must vary, cannot be determined without further data. It is well known that a similar cause produces in sound, that effect which is called a beat; two series of undulations of nearly equal magnitude cooperating and destroying each other alternately, as they coincide more or less perfectly in the times of performing their respective motions."

[37] *Ibid.* pp. 36-37.

[38] *Ibid.*, p. 45.

[39] *Phil. Trans.* 84 (1804), 7.

[40] *Ibid.*, p. 12.

[41] *Ibid.*, pp. 12-13.

[42] Using the corpuscular theory, the aberration angle could be derived by simply considering the components of velocity of light and the earth in orbit. The resultant produced the aberration angle.

[43] Grimaldi treats this phenomenon in Proposition xxii of his *Physico-Mathesis de Lumine, Coloribus et Iride* (Bologna, 1665). For a description of Grimaldi's work in optics, see Thomas Preston, *The Theory of Light* (London, 1890); Vasco Ronchi, *The Nature of Light;* A.I. Sabra, *Theories of Light from Descartes to Newton.*

[44] *Phil. Trans.* (1804), 11.

[45] *Ann. d. Phys.* 7 (1801), 527.

[46] *Phil. Trans.* 94 (1804), 15-16.

[47] *Ibid.*, p. 16.

[48] John F.W. Herschel, "Investigation of the Powers of the Prismatic Colours to heat and illuminate Objects; with Remarks that prove that the different Refrangibility of radiant Heat," *Phil. Trans.* 90 (1800), 255; "Experiments on the Refrangibility of the Invisible Rays of the Sun," *ibid.*, p. 284; "Experiments on the solar, and on the terrestrial Rays that occasion

Heat; with a comparative view of the laws to which light and Heat, or rather the rays which occasion them are subject in order to determine whether they are the same or different," *ibid.*, p. 293.

[49] Sir Henry Englefield, "Experiment on the Separation of Light and Heat by Refraction," *Nicholson Journal* (October 1802), 125.

[50] *Nicholson Journal* (July 1802), 181.

[51] Young published his explanation in a Note to Sir Henry's article, *Nicholson Journal* (July 1802), 183.

[52] "A Method of examining refractive and dispersive Powers by prismatic Reflection," *Phil. Trans.* 92 (1802), 365.

[53] *Phil. Trans.* 92 (1802), 381.

[54] *Ibid.*

[55] Wood, p. 168.

[56] Written by C. Innes in his Memoir of Dalzel. Dalzel, p. 214.

[57] Geoffrey Cantor has argued for placing Broughan within the Scottish methodological tradition. See G.N. Cantor, "Henry Brougham and the Scottish Methodological Tradition," *Studies in History and Philosophy of Science,* 2, No. 1 (May 1971), 69-89. Brougham is linked to the anti-conjectural tradition espoused by Thomas Reid. This linkage is persuasive, especially in terms of the rejection of the ether hypothesis. But I think Brougham's performance in his optical papers and his criticism of Young show him to be a naive follower of this tradition and an excellent illustration of the misapplication of the arguments presented. Brougham was the kind of disciple any master would have been happy to forego. For a further discussion of this tradition see L.L. Laudan, "Thomas Reid and the Newtonian Turn of British Methodological Thought" and Paul K. Feyerabend, "Classical Empiricism," both in *The Methodological Heritage of Newton,* ed. Robert E. Butts and John W. Davis. Both of these articles are more strongly critical of the tradition than is Cantor. Cantor's article discusses the Brougham-Young controversy in some detail. His emphasis is different from mine.

[58] *Edinburgh Review* I (1802-1803), 450-451.

[59] *Ibid.,* p. 452.

[60] *Edinburgh Review* V (1804-1805), 97.

[61] *Edinburgh Review* I (1802-1803), 99.

[62] *Ibid.,* V (1804), p. 400.

[63] Young's article "An Essay on Cycloidal Curves" was the third in a series of articles for the *British Magazine,* published under the name of the "Leptologist." *British Magazine,* 1 (1800), 314-320. Also reprinted in George Peacock, ed. *Miscellaneous Works of the Late Thomas Young,* I, p. 99. Brougham's paper was entitled "General Theorems, chiefly Porisms, in the higher Geometry," *Phil. Trans.* 88 (1798), 378-396.

[64] George Peacock, ed., *Misc. Works of the late Thomas Young,* I, p. 101. Hereafter cited as *Misc. Works.*

[65] *Misc. Works,* I, p. 102.

[66] *Phil. Trans.* 92 (1802), 35.

References

[67] *Ibid.*

[68] G.N. Cantor has argued that Brougham's motivations were primarily based upon methodology. He concluded his article "Henry Brougham and the Scottish Methodological Tradition" as follows: "In this reappraisal of the Young-Broughan controversy I have attempted to offer an alternative to the usual interpretation in which Brougham's reviews of Young have been attributed to personal animosity. Instead, I have tried to view the controversy as a dispute between two men holding very different attitudes toward science. In particular, Brougham rejected the wave theory of light because it involved a methodologically unpalatable hypothesis. Furthermore, I have tried to show that Brougham, in his rejection of hypotheses, in particular that of the luminiferous ether, was influenced by the Scottish methodology expounded by Thomas Reid. This case study is offered as an example of a scientific debate in which a methodological issue played a predominant role by preventing one of the disputants from appreciating the arguments in favor of a theory which was fairly well supported by empirical evidence." While I agree that it is important to link Brougham to his Edinburgh backgrounds, I think Cantor's article gives Brougham too much credit for both scientific competence and methodological concern. Brougham seemed to be more the converted believer to a dogma which must be protected from heresy by all means and at all costs. He grasped the letter and not the spirit of the Scottish methodological tradition, and used it to its worst advantage. Also, the element of nationalism entered Brougham's judgement. The opening years of the nineteenth century were not an appropriate time for Young to have cast doubt on a national hero, especially by reviving a theory derived from the work of Continental natural philosophers, particularly Frenchmen.

[69] Reprinted in *Misc. Works*, I, p. 192.

[70] *Misc. Works*, I, p. 192; Wood, p. 175.

[71] Wood, p. 175, taken from J.G. Lockhart, *Life of Sir Walter Scott* (London, 1896), p. 119.

[72] *Misc. Works*, I, p. 210; Wood, p. 173.

[73] Hudson Gurney, *A Memoir of the Life of Thomas Young*, p. 22.

[74] *A Course of Lectures on Natural Philosophy and the Mechanical Arts* (London, 1807) (2nd ed., London, 1845). These lectures were truly remarkable for their complete coverage and for the wide range of research indicated by extensive bibiographies appended to the lectures.

[75] Dated 29 March 1802. Dalzel, p. 212.

[76] Young certainly maintained his interest in optics. He produced a series of anonymous reviews for the *Quarterly Review* on articles recently appearing in optics. His reviews of papers by Laplace, Malus, and Biot are especially interesting. See *Miscellaneous Works of the Late Thomas Young*, ed. George Peacock for the reproduction of Young's reviews. See especially George Peacock's *Memoir of Dr. Thomas Young*, Chap. XII, for a description of Young's continued optical work and his relationship to the French, particularly Arago and Fresnel. The letters exchanged by Young, Arago, Laplace, and Fresnel are contained in *Misc. Works*, No. XVII, p. 359. See R.H. Silliman, *Augustin Fresnel and the Establishment of the Wave Theory of Light* (Princeton University, Ph.D. 1968) for a presentation of Young's work in optics, especially his relationship to Fresnel and the formulation of the concept of transverse waves in the ether. Silliman's assessment of the Young-Brougham controversy largely parallels my own. Edgar Morse, however, in his *Natural Philosophy, Hypotheses and Impiety: Sir David Brewster Confronts the Undulatory Theory of Light* (University of California, Berkeley, Ph.D. 1972) stresses the importance of Brougham's methodological concerns. He argues that Brougham's

criticisms were consistent in their attempt to protect the proper way to philosophize from Young's transgressions. Both Silliman and Cantor attribute more importance to Brougham's philosophical motivations than my own reading of the documents permits. I cannot view the controversy as hingeing upon scientific methodology, in part because of Brougham's choice of the new *Edinburgh Review* as his medium, in part because of my arguments as to how lightly Young held his notions of the ether, and in part because of Brougham's personality and personal interests. I agree that Brougham's general orientation and his defense of the tradition of "Bacon and Newton" should be seen to have some methodological content; see this chapter, footnote 68.

References

CHAPTER IV

[1] "Sur la Loi de la Réfraction Extraordinaire Dans Les Cristaux Diaphanes," *Journal de Physique* lxviii (1809), 107, and *Mémoires de Physique et de Chimie de la Société D'Arcueil*, Vol. II.

[2] William Herschel, as might be expected from his continued adherence to Newtonian optics, also chose to ignore Young's undulatory theory in his three papers on Newton's rings, *Phil. Trans.* 97 (1807), 180; 99 (1809), 259; and 100 (1810), 149. Herschel referred back to Henry Brougham's suggestion of the reflexity of the corpuscles of light and devised a substitute explanation for Newton's "Fits" in terms of color production by reflection. In his first paper Herschel rejected Newton's theory of fits: "But this, without mentioning particular objections, seems to be an hypothesis which cannot be easily reconciled with the minuteness and extreme velocity of the particles of which these rays, according to the *Newtonian* theory, are composed." Herschel rejected Newton's "Fits," neglected Young's law of interference, and performed his experiments on Newton's rings using two convex lenses which had belonged to Huygens.

[3] *Quarterly Review*, II (Nov. 1809), 337, also contained in *Misc. Works*, p. 220.

[4] See Vasco Ronchi, *The Nature of Light*, and Sir Edmund Whittaker, *A History of the Theories of Aether and Electricity* for a discussion of Malus's work. See Silliman, *Fresnel*, for an extended treatment of this period in the development of the wave theory.

[5] Article XV, "Memoirs de Physique et de Chemie de la Societe d'Arcueil," *Quarterly Review* (May 1810), 462. Young provided commentary on Malus' papers "On a Property of reflected Light" and "On a Property of the repulsive Forces which act on Light," pp. 472–477. Also in *Misc. Works*, pp. 246–253.

[6] *Ibid.*, p. 475.

[7] *Phil. Trans.* 103 (1813), 102.

[8] *Phil Trans.* 104 (1814), 219.

[9] *Ibid.*, p. 228.

[10] *Phil. Trans.* 104 (1814), 411.

[11] *Ibid.*, p. 415.

[12] *Phil. Trans.* 105 (1815), 149–450. I have not been able to determine the extent to which Brewster knew Biot's works at this time. His first mention of Biot's "Recherches sur la polarisation de la lumiere"and "Memoire sur la decouverte d'une propriete nouvelle dont jouissent les forces polarisantes de certain critaux" was in 1814. Brewster's knowledge of French science generally came from the reviews published in England. However, he did visit Paris in the summer of 1814, and probably met members of the French Academy. The members of the Academy were devoted to the corpuscular theory of optics, especially Biot, Laplace, and Poisson. Arago was the exception, and he was the champion of Fresnel's early papers.

[13] Edgar W. Morse provides a good short description of Brewster's scientific work, with a bibliography, in his article "David Brewster" for the *Dictionary of Scientific Biography*, Vol. II (New York: Charles Scribner's Sons, 1970), pp. 451–454.

[14] *Phil. Trans.* 105 (1815), 158-159.

[15] *Phil. Trans.* 106 (1816), 177.

[16] *Misc. Works,* p. 370. See pages 359-373 for correspondence between Young and Brewster.

[17] See Young's review article for a good summary of this work: "Review of Malus, Biot, Seebeck and Brewster on Light," *Quarterly Review* 9 (April 1814), 42.

[18] *Phil. Trans.* 108 (1818), 245.

[19] *Ibid.,* p. 246.

[20] *Ibid.,* p. 249.

[21] *Ibid.,* p. 267. "The increment of the square of the velocity of the extraordinary ray produced by the action of two axes of double refraction, is equal to the diagonal of a parallelogram whose sides are increments of the square of the velocity produced by each axis separately, and calculated by the law of Huygens' and whose angle is double and angle formed by the two planes passing through the ray and the respective axes."

[22] *Ibid.,* p. 270.

[23] *Phil. Trans.* 109 (1819), 158.

[24] The history of the development of the wave theory is a fascinating one and it remains to be completely written. For general descriptions of the wave theory see William Whewell, *History of the Inductive Sciences,* especially Book IX; Humphrey Lloyd, "Report on the Progress and Present State of Physical Optics," *British Association Reports* (1834), 295-415; M. Verdet's introduction to the edition of *Oeuvres Completes d'Augustin Fresnel* (Paris, 1866-70); Sir Edmund Whittaker, *A History of the Theories of Aether and Electricity,* Vol. I (London, 1910) (Harper Torchbook, 1960); Vasco Ronchi, *The Nature of Light (London, 1970).* See especially George Peacock, *Memoir of Dr. Thomas Young* and *Miscellaneous Works of the Late Thomas Young* (London, 1855) for letters between Young, Fresnel, and Arago on the developing wave theory. Peacock also describes Young's relationship to the French in Chap. XII. See especially two unpublished dissertations, Silliman, *Fresnel,* and D.B. Wilson, *The Reception of the Wave Theory of Light by Cambridge Physicists.*

[25] John Tyndall, *Six Lectures on Light* (London, 1873).

[26] *Phil. Trans.* 120 (1830), 133.

[27] *Ibid.,* p. 326.

[28] David Brewster, *A Treatise on Optics* (London: Longman, Green, 1831). The edition used here is the first American edition (Philadelphia, 1833), p. 118.

[29] See for example George Sarton, "Discovery of conical refraction by William Rowan Hamilton and Humphrey Lloyd (1833)," and D.B. Wilson, for the acceptance of the wave theory at Cambridge.

[30] "Report on the Progress and Present State of Physical Optics," By the Rev. Humphrey Lloyd, A.M., M.R.I.A., Fellow of Trinity College and Professor of Natural and Experimental Philosophy in the University of Dublin. *British Association Reports* (1834), 295-413.

[31] The history of the development of the wave theory of light in the nineteenth century has yet to be written. It is particularly important because optics was an important concern of most of the major figures in the physical sciences in the last half of the nineteenth century. Physical optics was a major component of the world view developed by "classical science."

Selected Bibliography

This bibliography has been selected to include only those primary and secondary works which have a bearing on the development of Newtonian optics. Those works by important contributors which do not directly relate to the topics have not been included. *Philosophical Transactions of the Royal Society* has been abbreviated throughout as *Phil. Trans.*

Alexander, H.G. *The Leibnitz-Clarke Correspondence.* (Manchester: University of Manchester Press, 1956).

Algarotti, Francesco. *The Lady's Philosophy.* (London, 1739).

Austin, William H. "Isaac Newton on Science and Religion," *Journal of the History of Ideas* 31 (1970), 521–542.

Badcock, A.W. "Physical Optics at the Royal Society," *The British Journal for the History of Science* 1 (1962), 99–116.

Ball, W.W. Rouse. *A History of the Study of Mathematics at Cambridge.* (Cambridge: Cambridge University Press, 1889).

Bell, A.E. *Christian Huygens and the Development of Science in the Seventeenth Century.* (New York: Longmans Green & Co., 1947).

———. *Newtonian Science.* (London: Edward Arnold Ltd., 1961).

Bentley, Richard. *Works of Richard Bentley.* (London, 1838).

Boscovich, Roger Joseph. *A Theory of Natural Philosophy*, Latin-English edition. (Chicago: Open Court, 1922).

Boyle, Hon. Robert. *Experiments and considerations touching colours.* (London: Henry Herringman, 1670).

Brewster, Sir David. *The Life of Sir Isaac Newton.* (New York: J. and J. Harper, 1831).

———. *A Treatise on New Philosophical Instruments for various purposes in the Arts and Sciences with experiments on light and colours.* (Edinburgh, 1813).

———. *A Treatise on Optics.* (Philadelphia, 1833).

———. *Letters of Euler on Different Subjects in Natural Philosophy Addressed to a German Princess with Notes and a life of Euler by David Brewster.* (New York: J. and J. Harper, 1833).

———. "On Some Properties of Light," *Phil. Trans.* (1813), 101.

———. "On the Affections of Light transmitted through crystallized Bodies," *Phil. Trans.* (1814), 187.

———. "On the Polarization of Light by oblique transmission through all bodies, whether crystallized or uncrystallized," *Phil. Trans.* (1814), 219.

———. "On new Properties of light exhibited in the optical Phenomena of Mother of Pearl, and other Bodies to which the superficial structure of that Substance can be communicated," *Phil. Trans.* (1814), 397.

———. "Additional observations on the optical properties and structure of heated glass and unannealed glass drops," *Phil. Trans.* (1815), 1.

———. "Experiments on the depolarisation of Light as exhibited by various mineral, animal, and vegetable bodies, with a reference of the phenomena to the general principles of polarisation," *Phil. Trans.* (1815), 29.

———. "On the laws which regulate the polarisation of light by reflexion from transparent bodies," *Phil. Trans.* (1815), 125.

———. "On the multiplication of images, and the colours which accompany them in some specimens of calcareous spar," *Phil. Trans.* (1815), 270.

———. "On New properties of heat, as exhibited in its propagation along plates of glass," *Phil. Trans.* (1816), 46.

———. "On the communication of the structure of doubly refracting crystals to glass, muriate of soda, fluor spar, and other substances, by mechanical compression and dilation," *Phil. Trans.* (1816), 156.

———. "On the structure of the crystalline lens in fishes and quadrupeds, as ascertained by its action on polarized light," *Phil. Trans.* (1816), 311.

———. "On the laws of polarization and double refraction in regularly crystallized bodies," *Phil. Trans.* (1818), 199.

———. "On the Laws which regulate the Absorption of polarised light by Double Refracting Crystals," *Phil. Trans.* (1819), 11.

———. "On the action of Crystallized surfaces upon Light," *Phil. Trans.* (1819), 145.

———. "On a new series of periodical colours produced by the grooved surfaces of metallic and transparent bodies," *Phil. Trans.* (1829), 301.

———. "On the law of the partial polarization of light by reflexion," *Phil. Trans.* (1830), 69.

———. "On the production of double refraction in the molecules of bodies by simple pressure; with observations on the origin of the doubly refracting structure," *Phil. Trans.* (1830), 87.

———. "On the laws of polarization of light by refraction," *Phil. Trans.* (1830), 133.

———. "On the action of the second surfaces of transparent plates upon light," *Phil. Trans.* (1830), 145.

———. "On the Phenomena and Laws of Elliptic Polarization as exhibited in the Action of Metals upon Light," *Phil. Trans.* (1830), 287.

———. "On the Colours of Mixed Plates," *Phil. Trans.* (1838), 73.

Brougham, Henry. "Experiments and Observations on the Inflection, Reflection, and Colours of Light," *Phil. Trans.* 86 (1796), 227.

———. "Farther Experiments and Observations on the Affections and Properties of Light," *Phil. Trans.* 87 (1797), 352.

Brunet, Pierre. *L'Introduction des Theories de Newton en France au XVIIIe siecle.* (Paris: Libraire Scientifique Albert Blanchard, 1931).

Buchdahl, Gerd. "Gravity and Intelligibility: Newton to Kant," in Butts, R.E., and Davis, J.W. (eds.), *The Methodological Heritage of Newton.* (Toronto: The University of Toronto Press, 1970.)

Burtt, E.A. *The Metaphysical Foundations of Modern Science.* (New York: Doubleday, 1932).

Butts, Robert E., and Davis, John W. (eds.). *The Methodological Heritage of Newton.* (Toronto: University of Toronto Press, 1970).

Byrom, John. *The Private Journal and Literary Remains of John Byrom.* Edited by Richard Parkinson. (London: Printed for the Chetham Society, 1854).

Cohen, I. Bernard. "The First Explanation of Interference," *American Journal of Physics* 8 (1940), 99.

———. (ed.). *Isaac Newton's Papers and Letters on Natural Philosophy.* (Cambridge: Harvard University Press, 1958).

———. *Franklin and Newton.* (Cambridge: Harvard University Press, 1966).

Courtney, William Prideaux. "Robert Smith," *Dictionary of National Biography*, Vol. 18. (London, 1917), p. 518.

Crew, Henry. *The Wave Theory of Light, Memoirs of Huygens, Young and Fresnel.* (New York: American Book Company, 1900).

Cumberland, Richard. *Memoirs of Richard Cumberland written by himself.* Vol. I and II. (London: Lockington, Allen and Company, 1807).

Dalzel, Andrew. *History of the University of Edinburgh from its first Foundation* with a memoir of the Author by C. Innes. (Edinburgh: Edmonston and Douglas, 1862).

Dijksterhuis, E.J. *The Mechanization of the World Picture.* (Oxford: Clarendon Press, 1961).

Dolland, John. "A Letter from Mr. John Dolland to James Short, concerning a mistake in M. Euler's Theorem for correcting the Aberrations in the Object Glasses of refracting Telescopes," *Phil. Trans.* 48 (1752), 289.

———. "An Account of some Experiments Concerning the different Refrangibility of Light," *Phil. Trans.* 50 (1754), 733.

Englefield, Sir Henry. "On the Effect of Sound upon the Barometer" with a "Note By Dr. Young," *A Journal of Natural Philosophy, Chemistry and the Arts* (July, 1802), 181.

———. "Experiments on the Separation of Light and Heat by Refraction, In a Letter from Sir H.C. Englefield, to Thomas Young," *A Journal of Natural Philosophy, Chemistry and the Arts* (October, 1802), 125.

Euler, Leonhard. *Letters of Euler on different subjects in Physics and Philosophy addressed to a German Princess.* Translated with notes by Henry Hunter. (London, 1802).

———. "Memoire sur l'effet de la propagation successive de la lumiere dans l'apparition tant des planetes que des cometes," *Histoire de l'Academie Royale des Sciences et Belles-Lettres [de Berlin] avec les memoires* 2 (1746).

———. *Nova theoria lucis et colorum,* in *Opuscula varii argumenti,* 1 (1746), p. 169 (Berlin).

———. "Recherches physiques sur la cause des queues de cometes, de la lumiere Boreale et de la lumiere zodiacle," *Histoire de l'Academie Royale des Sciences et Belles-Lettres [de Berlin] avec les memoires* 2 (1746).

———. "Sur la perfection des verres objectifs des lunettes," *Histoire de l'Academie Royale des Sciences et Belles-Lettres [de Berlin] avec les memoires* 3 (1747), 274-296.

———. *Conjectura physica de propagatione soni ac luminis,* in *Opuscula varii argumenti* 2 (1750), p. 1 (Berlin).

———. "Essai d'une explication physique des couleurs engendrees sur des surfaces extremement minces," *Histoire de l'Academie Royale des Sciences et Belles-Lettres [de Berlin] avec les memoires* 8 (1752).

———. "Examen d'une controverse sur la loi de refraction des rayons de differentes couleurs par rapport a la diversite des milieux," *Histoire de l'Academie Royale des Sciences et Belles-Lettres [de Berlin] avec les memoires* 9 (1753).

———. "Letter from M. Euler to Mr. Dolland," *Phil. Trans.* 48 (1753), 292.

———. "Letter from Mr. Euler to Mr. James Short, F.R.S.," *Phil. Trans.* 48 (1753), 292.

———. "Recherches physiques sur la diverse refrangibilite des rayons de lumiere," *Histoire de l'Academie Royale des Sciences et Belles-Lettres [de Berlin] avec les memoires* 10 (1754).

Feyerabend, Paul K. "Classical Empiricism," in Butts, R.E., and Davis, J.W. (eds.), *The Methodological Heritage of Newton.* (Toronto: University of Toronto Press, 1970).

Foote, George A. "Mechanism, Materialism and Science in England, 1800-1850," *Annals of Science* 8 (1952).

Fuss, P.H. *Correspondance Mathematique et Physique de Quelques Celebres Geometres Du XVIIIeme Siecle,* Tomes I et II. (St. Petersbourg, 1843).

Gough, John. "Reply to Dr. Young's Letter on the Theory of Compound Sounds," *A Journal of Natural Philosophy, Chemistry and the Arts* (November, 1802).

Gregory, David. *Elements of catoptrics and dioptrics,* Translated from the Latin original, with a large supplement by W. Browne. (London, 1735).

Guerlac, Henry. "Three Eighteenth-Century Social Philosophers: Scientific Influences on Their Thought," in Holton, Gerald (ed.), *Science and the Modern Mind.* (Boston: Beacon Press, 1958), pp. 1-18. First published in *Daedalus.*

———. "Sir Isaac Newton and the Ingenious Mr. Hawksbee," *L'aventure de la Science, Melanges Alexandre Koyré* I (Paris: Herman, 1964), p. 228.

Selected Bibliography

———. "Where the Statue Stood: Divergent Loyalties to Newton in the Eighteenth Century," in Wasserman, Earl Reeves (ed.), *Aspects of the Eighteenth Century*. (Baltimore: The Johns Hopkins Press, 1965), pp. 317-334.

———. "Newton's Optical Aether, His Draft of a Proposed Addition to his *Opticks*," *Notes and Records of the Royal Society of London* 22 (1967), 45-57.

Gurney, Hudson. *Memoir of the Life of Thomas Young*. (London: John and Arthor Arch, 1831).

Hall, A.R. "Sir Isaac Newton's Note-Book, 1661-65," *Cambridge Historical Journal* 9 (1948), 239.

——— and Hall, M.B. "Newton's Electric Spirit: Four Oddities," *ISIS* 50 (1959), 473-476.

Hall, A. Rupert and Marie Boas. *Unpublished Scientific Papers of Sir Isaac Newton*. (Cambridge, 1962).

Hall, A. Rupert. *From Galileo to Newton, 1630-1720*. (New York: Harper & Row, 1963).

Hardin, Clyde L. "The Scientific Work of the Reverend John Michell," *Annals of Science* 22 (1966), 27-47.

Hardy, W.B. "Historical Notes upon Surface Energy and Forces of Short Range," *Nature* 109, No. 2734 (March 23, 1922), 375-378.

Harris, John. *Lexicon Technicum*. (London: D. Brown, 1704).

Helsham, Richard. *A Course of Lectures in Natural Philosophy*. (London: Published by Brian Robinson, 1739).

Herschel, Sir John. *Familiar Lectures on Scientific Subjects*. (London: Alexander Strahan and Co., 1866).

———. "On the action of crystallized bodies on homogeneous light, and on the causes of the deviation from Newton's scale in the tints which many of them develop on exposure to a polarized ray," *Phil. Trans.* 110 (1820), 45.

Herschel, William. *The Scientific Papers of Sir William Herschel*, including early papers hitherto unpublished. With biographical intro. by J.L.E. Dreyer. 2 Vols. (London, 1912).

———. "Investigation of the Powers of the Prismatic colours to heat and illuminate Objects; with Remarks that prove the different Refrangibility of radiant Heat," *Phil. Trans.* (1800), 255.

———. "Experiments on the Refrangibility of the Invisible Rays of the Sun," *Phil. Trans.* (1800), 284.

———. "Experiments on the solar, and on the terrestrial Rays that occasion Heat; with a comparative view of the laws to which light and Heat, or rather the rays which occasion them, are subject in order to determine whether they are the same or different," *Phil. Trans.* (1800), 293.

———. "Experiments for investigating the Cause of the coloured concentric Rings, discovered by Sir Isaac Newton, between two Object-glasses laid upon one another," Read February 5, 1807. *Phil. Trans.* (1807), 180-233.

———. "Continuation of Experiments for investigating the Cause of coloured Concentric Rings, and other Appearances of a similar Nature," Read March 23, 1809. *Phil. Trans.* (1809), 259-302.

———. "Supplement to the First and Second Part of the Paper of Experiments, for Investigating the Cause of Coloured Concentric Rings between Object Glasses, and other Appearances of a similar Nature," Read March 15, 1810. *Phil. Trans.* (1810), 149-177.

Hesse, Mary B. *Forces and Fields*, a study of action at a distance in the history of physics. Originally published 1961. (New Jersey: Littlefield, Adams & Co., 1965).

Higgins, Brian. *A Philosophical Essay Concerning Light*. (London: Printed for J. Dodsley, in Pall Mall, 1776).

Holton, Gerald (ed.). *Science and the Modern Mind*, A Symposium. (Boston: Beacon Press, 1958).

Hooke, Robert. *Micrographia, or Some physiological Descriptions of Minute Bodies made by Magnifying Glasses with Observations and Inquiries thereupon*. Facsimile reproduction of the first edition, Royal Society, 1665. (New York: Dover, Inc., 1961).

Hoppe, E. *Geschichte der Optik*. (Leipzig, 1926).

Huxley, George. "Roger Cotes and Natural Philosophy," *Scripta Mathematica* 26 (1961), 231-238.

Huygens, Christian. *Oeuvres Completes*. (The Hague, 1888-1950).

———. *Treatise on Light*, trans. by Silvanus P. Thompson. (Dover Publications, 1962).

Kargon, Robert Hugh. *Atomism in England from Hariot to Newton*. (Oxford: Clarendon Press, 1966).

Keill, John. *An Introduction to Natural Philosophy*. (London, 1758).

Knight, D.M. "The Physical Sciences and the Romantic Movement," *History of Science* 9 (1970), 54-73.

Knox, John. "On Some phenomena of colours, exhibited by thin plates," *Phil. Trans.* 105 (1815), 161.

Koyré, Alexander and Cohen, I. Bernard. "Newton and the Leibnitz-Clarke Correspondence," *Archives Internationales d'histoire des Sciences* Nos. 58-59 (1962), 63-126.

———. *Newtonian Studies*. (Chicago: Phoenix Books, 1968).

Laudan, L.L. "Thomas Reid and the Newtonian Turn of British Methodological Thought," in Butts, R.E., and Davis, J.W. (eds.), *The Methodological Heritage of Newton*. (Toronto: University of Toronto Press, 1970).

Layton, David. "Diction and Dictionaries in the Diffusion of Scientific Knowledge: An Aspect of the History of the Popularization of Science in Great Britain," *Brit. J. Phil. Sci.* 2 (1964/5), 221-234.

Selected Bibliography

Larmor, Sir Joseph. "Thomas Young," *Nature* 133, No. 3356 (February 24, 1934).

Lloyd, Humphrey. "Report on the Progress and Present State of Physical Optics," *British Association Reports* (1834), 295-415.

McCormmach, Russell. "John Michell and Henry Cavendish: Weighing the Stars," *Brit. J. Hist. Sci.* 4 (1968), 126-155.

―――――. "Henry Cavendish: A Study of Rational Empiricism in Eighteenth-Century Natural Philosophy," *ISIS* 60, No. 203 (1969), 293-306.

McGuire, J.E. and Rattansi, P.M. "Newton and the 'Pipes of Pan,'" *Notes and Records of the Royal Society of London.* 21, No. 2 (December 1966), 108-143.

McGuire, J.E. "Body and Void and Newton's De Mundi Systemate: Some New Sources," *Archive for History of Exact Sciences* 3 (1966/67), 208-248.

―――――. "The Origin of Newton's Doctrine of Essential Qualities," *Centaurus* 12, No. 4 (1968), 233-260.

Mach, Ernst. *The Principles of Physical Optics.* (New York: Dover Publications, 1926).

Maclaurin, Colin. *An Account of Sir Isaac Newton's Philosophical Discoveries.* (London, 1748).

Manuel, Frank. "Newton as Autocrat of Science," *Daedalus* 97 (1968), 969-1001.

Marsak, Leonard M. "Cartesianism in Fontenelle and French Science, 1686-1752," *ISIS* 50 (1959), 51-60.

Martin, Benjamin. *The Philosophical Grammar.* (London, 1738).

―――――. *A New and Compendious System of Optics.* (London: James Hodges, 1740).

―――――. *A Plain and Familiar Introduction to the Newtonian Philosophy in Six Lectures.* (London: W. Owen, 1751).

―――――. *New Elements of Optics.* (London, 1759).

Melvil, Thomas. "A Letter from Mr. T. Melvil to the Rev. James Bradley, D.D.F.R.S. With a Discourse concerning the Cause of the different Refrangibility of the Rays of Light," *Phil. Trans.* 48 (1753), 261-269.

―――――. "Observations on Light and Colours," *Essays and Observations Physical and Literary* of the Royal Philosophical Society of Edinburgh (1770-1).

Merz, John Theodore. *A History of European Thought in the 19th Century.* (Edinburgh: William Blackwood, 1903).

Michell, John. "On the means of discovering the distance, magnitude, etc. of the fixed stars . . . ," *Phil. Trans.* 74 (1784), 35-57.

Molyneux, William. *Dioptrica Nova.* (London, 1692).

Morse, E.W. *Natural Philosophy, Hypotheses and Impiety: Sir David Brewster Confronts the Undulatory Theory of Light.* University of California, Berkeley, Ph.D. 1972.

Murdock, Patrick. "Rules and Examples for limiting the Cases in which the Rays of refracted Light may be reunited into a colourless Pencil," *Phil. Trans.* 53 (1758), 173.

Newton, Sir Isaac. *Opticks or a Treatise of the Reflections, Refractions, inflections and colours of light.* (New York: Dover Publications, 1952).

———. *Mathematical Principles of Natural Philosophy and his System of the World*. Andrew Motte's translation revised by Florian Cajori. (Berkeley: University of California Press, 1934).

Nicolson, Marjorie Hope. *Newton Demands the Muse, Newton's Opticks and the 18th Century Poets*. (Princeton: Princeton University Press, 1946).

Pav, Peter Anton. "Eighteenth Century Optics: The Age of Unenlightenment," Indiana University Ph.D. dissertation, 1964. (University Microfilms, Inc., Ann Arbor, Michigan, No. 65-3510).

Peacock, George. *Life of Thomas Young*. (London: John Murray, 1855).

———. (ed.). *Miscellaneous Works of the late Thomas Young*. Vol. I and II ed. G. Peacock, Vol. III ed. J. Leitch. (London, 1858).

Pemberton, Henry. *A View of Sir Isaac Newton's Philosophy*. (London: S. Palmer, 1728).

Perrole. "A Philosophical Memoir, containing (1) Experiments relative to the Propagation of Sound in different Solid and Fluid Mediums, and (2) An Experimental Inquiry into the Cause of the Resonance of Musical Instruments," *A Journal of Natural Philosophy, Chemistry and the Arts* I (1797).

Pettigrew, Joseph. *Medical Portrait Gallery. Biographical Memoirs of the most Celebrated Physicians, Surgeons etc., etc. who have contributed to the Advancement of Medical Science*. (London: Whittaker & Co., n.d.)

Priestley, Joseph. *The History and Present State of Discoveries relating to vision, light and colours*. (London: J. Johnson, 1772).

Roberts, M., and Thomas, E.R. *Newton and the Origin of Colours*. (London: G. Bell and Sons, 1934).

Robison, John, "On the Motion of Light, as affected by refracting and reflecting Substances, which are also in Motion," *Trans. Royal Soc. of Edin.* 2 (1788), 83-111.

Robinson, Bryan. *A Dissertation on the Aether of Sir Isaac Newton*. (Dublin, 1743).

———. *Sir Isaac Newton's Account of the Aether with Some Additions by Way of Appendix*. (Dublin, 1745).

Ronchi, Vasco. *The Nature of Light*, an Historical Survey. Trans. V. Barocas. Originally published as *Storia della Luce*, 1939. (London: Heinemann, 1970).

Rowning, John. *Compendious System of Natural Philosophy*. (London: Sam Harding, 1738).

Rutherford, T. *A System of Natural Philosophy being a course of lectures in Mechanics, Optics, Hydrostatics and Astronomy*. (Cambridge: J. Bentham, 1748).

Sabra, A.I. *Theories of Light from Descartes to Newton*. (London: Oldbourne Book Co., Ltd., 1967).

Sarton, George. "Discovery of Conical Refraction by William Rowan Hamilton and Humphrey Lloyd (1833)," *ISIS* 17 (1932), 154-170.

Schofield, Robert E. "Joseph Priestley, Natural Philosopher," *Ambix* 14 (1967), 1-15.

———. *Mechanism and Materialism*, British Natural Philosophy in an Age of Reason. (Princeton: Princeton University Press, 1970).

Silliman, R.H. *Augustin Fresnel and the Establishment of the Wave Theory of Light*, Princeton University, Ph.D., 1968.

Smith, Robert. *A Compleat System of Optics in 4 Books, viz. a popular, a mathematical, a mechanical and a philosophical treatise*. (Cambridge: Printed for the Author, 1738).

———. *Harmonics, or the Philosophy of musical sounds*. (Cambridge: Printed by J. Bentham and sold by W. Thurlbaum, 1749).

———. *Harmonics, or the Philosophy of musical sounds. 2nd Edition much improved and augmented*. (London: Printed for T. and J. Merrill, 1759).

———. *A Postscript to Dr. Smith's Harmonics*. (Cambridge: Printed for T. and J. Merrill, 1762).

Thomson, Benjamin, Count of Rumford. "An Inquiry concerning the Nature of Heat, and the mode of its communication," *Phil. Trans.* (1804); *Edinburgh Review*, Article XI (July 1804).

Tyndall, John. *Six Lectures on Light, delivered in America in 1872-1873*. (London: Longmans, Green and Company, 1873).

Vince, Rev. Samuel. The Bakerian Lecture, "Observations on the Theory of the Motion and Resistance of Fluids: with a Description of the Construction of Experiments, in order to obtain some fundamental Principles," *Phil. Trans.* 84 (1794).

Westfall, Richard. "The Development of Newton's Theory of Color," *ISIS* 53 (1962), 339.

———. "Newton's Optics: The Present State of Research," *ISIS* 57 (1966), 102-107.

Whewell, William. *History of the Inductive Sciences, from earliest to the Present Times*. 3rd ed. (New York: D. Appleton and Company, 1894).

Whittaker, Edmund. *A History of the Theories of Aether and Electricity*. Vol. I. (New York: Harper Torchbook, 1960).

Whyte, Lancelot Law. *Roger Joseph Boscovich*. (London: G. Allen and Unwin, 1961).

Wilson, D.B. *The Reception of the Wave Theory of Light by Cambridge Physicists*, The Johns Hopkins University, Ph.D., 1968.

Wightman, William P.D. *The Growth of Scientific Ideas*. (New Haven: Yale Press, 1964).

Wolf, A. *A History of Science, Technology and Philosophy in the 18th Century*. (New York: Harper Torchbook, 1961).

Wollaston, William Hyde. "A Method of examining refractive and dispersive Powers by prismatic Reflection," *Phil. Trans.* 92 (1802), 362.

———. "On the oblique Refraction of Iceland Crystal," *Phil. Trans.* 92 (1802), 381.

———. "A Method of Examining Refractive and Dispersive Powers by Prismatic Reflection," *Phil. Trans.* (1802); *Edinburgh Review*, Article VIII (April 1803).

———. "On the Oblique Reflection of Iceland Crystal," *Phil. Trans.* (1802); *Edinburgh Review*, Article IX (April 1803).

Wood, Alexander. *Thomas Young, Natural Philosopher.* Completed by Frank Oldham for the author. (Cambridge: University Press, 1954).

Wood, James. "The Elements of Optics designed for the use of Students of the University," (Cambridge, 1801), *Edinburgh Review*, Article XXIII (October 1802).

Young, Thomas. *A Course of Lectures on Natural Philosophy and the Mechanical Arts.* 2 Vols. (London: Joseph Johnson, 1807).

———. *A Course of Lectures on Natural Philosophy and the Mechanical Arts.* A New Edition, with References and Notes by Rev. P. Kelland. (London: Printed for Taylor & Walton, 1845).

———. "Outlines of Experiments and Inquiries respecting Sound and Light," *Phil. Trans.* 90 (1800).

———. The Bakerian Lecture, "On the Mechanism of the Eye," *Phil. Trans.* 91 (1801).

———. "A Letter from Thomas Young, M.D., F.R.S. Professor of Natural Philosophy in the Royal Institution, respecting Sound and Light and in Reply to some Observations of Professor Robison," *A Journal of Natural Philosophy, Chemistry and the Arts* (August 1801).

———. "An Answer to Mr. Gough's Essay on the Theory of Compound Sounds," *A Journal of Natural Philosophy, Chemistry and the Arts* (August 1802).

———. "A Syllabus of a Course of Lectures on Natural and Experimental Philosophy" (Book Review), *A Journal of Natural Philosophy, Chemistry and the Arts* (March 1802), 239.

———. The Bakerian Lecture, "On the Theory of Light and Colours," *Phil. Trans.* 92 (1802).

———. "The Bakerian Lecture on the Theory of Light and Colours," *Phil. Trans.* (1802); *Edinburgh Review*, Article XVI (January 1803).

———. "An Account of some Cases of the Production of Colours, not Hitherto described," *Phil. Trans.* 92 (1802), 387.

———. "A Summary of the Most useful Parts of Hydraulics, chiefly extracted and abridged from Eytelwein's Handbuch der Mechanik und der Hydraulik (Berlin, 1801)," *A Journal of Natural Philosophy, Chemistry and the Arts* (September 1802), 25.

———. "Letter from Thomas Young, M.D.F.R.S. In Reply to Mr. Gough's Letter, at Page 36 of the present Volume On the Phenomena of Sound," *A Journal of Natural Philosophy, Chemistry and the Arts* (November 1802), 145.

Selected Bibliography

———. The Bakerian Lecture, "Experiments and Calculations relative to physical Optics," *Phil. Trans.* 94 (1804).

———. "The Bakerian Lecture, Experiments and Calculations relative to Physical Optics," *Phil. Trans.* (1804); *Edinburgh Review*, Article VII (October 1804).

———. "Review of Laplace's Memoir Sur la Loi de la Refraction Extraordinaire dans les Cristaux Diaphanes," *Quarterly Review* ii (November 1809), 337.

———. "Review of the Memoires de Physique et de Chimie de la Societe D'Arcueil, vols. I and II," *Quarterly Review* iii (May 1810), 462.

———. "Review of Malus, Biot, Seebeck, and Brewster on Light," *Quarterly Review* XI (April 1814), 42.

Index

A Compleat System of Opticks, 27-28, 34-47, 50, 71, 83
Atomism, 3-5, 23-24

Bacon, Francis, 9
Barrow, Isaac, 5, 11
Bentley, Richard, 14, 21, 32
Biot, 144-145
Boscovich, Roger Joseph, 71-73, 80, 85
Boyle, Robert, 5, 10-11, 24
Brewster, David, Chapter IV
Brougham, Henry, 86-92, 117, 122, 128-136, 141

Cambridge Platonists, 11, 17
Cavendish, Henry, 67-68, 75-80
Clarke, Samuel, 29
Color theory, 43-45, 92-94, 103-104, 121-125, 131-132
Comet's tails, 15-16
Corpuscular dynamics, 48
Cotes, Roger, 21, 27-34, 100, 102

De Mairan, Dortous, 65
Desaguliers, J.T., 50
Descartes, René, 6, 9-10, 93-95
Dolland, John, 55-59, 64, 77

Edinburgh Review, 128
Englefield, Sir Henry, 126
Ether, 46-48, 54, 92-95, 102, 112-113, 120-124, 150
Euler, Leonhard, 55-57, 70, 72, 102-105, 113-114
Eye, 108

Fits, 44-47, 62-64, 73, 91-92, 97-98, 114, 148-149
Foucault, Jean Leon, 150
Franklin, Benjamin, 70
Freind, John, 48
Fresnel, Auguste Jean, 147-148

Gassendi, Pierre, 6-7, 9
Gough, John, 119

Grimaldi, 100, 124

Halley, Edmund, 100, 116
Harmonics, 109-110, 114-115, 118-119
Harriot, Thomas, 5-6
Harris, John, 48-50
Hauksbee, Francis, 42
Heat, 125-126, 130
Helsham, Richard, 53
Herschel, William, 76, 80-83, 125-126
Higgins, Brian, 104-106
Hobbes, Thomas, 6-10
Hooke, Robert, 92-94, 116
Huygens, Christian, 92, 94-102, 113

Interference, 121-125, 131-132

Keil, John, 48
Kepler, Johannes, 6
Klingenstierna, Samuel, 56

Laplace, Pierre Simon, 137-138, 144
Lloyd, Humphrey, 150
Luminescence, 47

Malebranche, 103-104
Malus, Etienne Louis, 138-139
Martin, Benjamin, 51-53, 107
Maskelyne, Nevil, 77
Materialism, 54, 66-67
Mechanical philosophy, 6, 17-18, 24-25
Mechanist interpretation, 65-67
Melvil, Thomas, 60-63
Michell, John, 67-80
Molyneux, William, 49, 51
Momentum of light, 64-65, 72
More, Henry, 11-13

Newton, 1-27
  and Huygens, 97-102
  and Newtonians, 1-2, 8-9, 13, 19, 26-27, 31, 101-102
  concept of God, 13-17, 20-22, 32

Opticks, 22-23, 34-35, 38-39, 41-42, 44, 56-57, 79, 87, 101, 107

Pemberton, Henry, 50, 63

Playfair, John, 85
Polarization, 138-140, 142-148
Priestley, Joseph, 59, 64-65, 68-74, 82-83
Principia, 36-38, 79, 88, 97
Prisca tradition, 17, 19

Queries, 20-23, 25, 34-35, 47, 50

Ritter, J.W., 125
Robison, John, 83-86, 118
Rowning, John, 53
Rumford, Count, 126, 130
Rutherford, T., 53

Short, James, 63-64
Smith, Robert, 27-28, 33-48, 50, 71, 83, 109-110, 114-115, 118-119
Sound, 109-112, 126

Tides, 116-117
Tyndall, John, 147

Wollaston, William Hyde, 126-127, 130

Young, Thomas, Chapter III, 141
and Newton, 111-112, 120-122

Zones of force, 39, 44-47, 80-82, 86-89